Lecture Notes in Computer Science

Commenced Publication in 1973
Founding and Former Series Editors:
Gerhard Goos, Juris Hartmanis, and Jan van Leeuwen

Yishai A. Feldman
Donald Kraft
Tsvi Kuflik (Eds.)

Next Generation Information Technologies and Systems

7th International Conference, NGITS 2009
Haifa, Israel, June 16-18, 2009
Revised Selected Papers

 Springer

Volume Editors

Yishai A. Feldman
IBM Haifa Research Lab
Haifa University Campus, Mount Carmel, Haifa 31905, Israel
E-mail: yishai@il.ibm.com

Donald Kraft
U.S. Air Force Academy
Department of Computer Science
2354 Fairchild Drive, Suite 6G-101, Colorado Springs, CO 80840, USA
E-mail: donald.kraft@usafa.edu

Tsvi Kuflik
The University of Haifa
Management Information Systems Department
Mount Carmel, Haifa 31905, Israel
E-mail: tsvikak@is.haifa.ac.il

Library of Congress Control Number: 2009935905

CR Subject Classification (1998): H.4, H.3, H.5, H.2, D.2.12, C.2.4

LNCS Sublibrary: SL 3 – Information Systems and Application, incl. Internet/Web
and HCI

ISSN   0302-9743
ISBN-10   3-642-04940-0 Springer Berlin Heidelberg New York
ISBN-13   978-3-642-04940-8 Springer Berlin Heidelberg New York

springer.com

© Springer-Verlag Berlin Heidelberg 2009
Printed in Germany

Typesetting: Camera-ready by author, data conversion by Scientific Publishing Services, Chennai, India
Printed on acid-free paper   SPIN: 12772971   06/3180   5 4 3 2 1 0

Foreword

Information technology is a rapidly changing field in which researchers and developers must continuously set their vision on the next generation of technologies and the systems that they enable. The Next Generation Information Technologies and Systems (NGITS) series of conferences provides a forum for presenting and discussing the latest advances in information technology. NGITS conferences are international events held in Israel; previous conferences have taken place in 1993, 1995, 1997, 1999, 2002, and 2006.

In addition to 14 reviewed papers, the conference featured two keynote lectures and an invited talk by notable experts. The selected papers may be classified roughly in five broad areas:

- Middleware and Integration
- Modeling
- Healthcare/Biomedical
- Service and Information Management
- Applications

NGITS 2009 also included a demonstration session and an industrial track focusing on how to make software development more efficient by cutting expenses with technology and infrastructures.

This event is the culmination of efforts by many talented and dedicated individuals. We are pleased to extend our thanks to the authors of all submitted papers, the members of the program committee, and the external reviewers. Many thanks are also due to Nilly Schnapp for local organization and logistics, and to Eugeny Myunster for managing the web site and all other technical things. Finally, we are pleased to acknowledge the support of our institutional sponsors: The University of Haifa, the Faculty of Social Sciences and the MIS Department at the University of Haifa, the IBM Haifa Research Lab, and the Technion.

June 2009

Yishai A. Feldman
Donald Kraft
Tsvi Kuflik

Organization

General Chair

Tsvi Kuflik

Steering Committee

Opher Etzion   Avigdor Gal   Amihai Motro

Program Committee Chairs

Yishai A. Feldman   Donald Kraft

Program Committee

Nabil Adam
Hamideh Afsarmanesh
Mathias Bauer
Iris Berger
Dan Berry
Elisa Bertino
Gloria Bordogna
Patrick Bosc
Rebecca Cathey
Jen-Yao Chung
Alessandro D'Atri
Asuman Dogac
Ophir Frieder
Mati Golani
Paolo Giorgini
Enrique Herrera-Viedma
David Konopnicki

Manolis Koubarakis
Maria Jose Martin-Bautista
Amnon Meisels
Naftaly Minsky
George Papadopoulos
Gabriella Pasi
Mor Peleg
Haggai Roitman
Doron Rotem
Steve Schach
Pnina Soffer
Bracha Shapira
Bernhard Thalheim
Eran Toch
Yair Wand
Ouri Wolfson
Amiram Yehudai

Additional Reviewers

Gunes Aluc
Joel Booth
Alessio Maria Braccini
Mariangela Contenti
Ekatarina Ermilova

Gokce Banu Laleci Erturkmen
Stefania Marrara
Simon Samwel Msanjila
Cagdas Ocalan
Aabhas Paliwal

Andrea Resca   Venkatakumar Srinivasan
Basit Shafiq   Fulya Tuncer
Michal Shmueli-Scheuer   Stefano Za

Local Arrangements

Nilly Schnapp

Website Manager

Eugeny Myunster

Table of Contents

4   Service and Information Management

5   Applications

Searching in the "Real World"
(Abstract of Invited Plenary Talk)

Ophir Frieder

Information Retrieval Laboratory
Department of Computer Science
Illinois Institute of Technology
ophir@ir.iit.edu

For many, "searching" is considered a mostly solved problem. In fact, for text processing, this belief is factually based. The problem is that most "real world" search applications involve "complex documents", and such applications are far from solved. Complex documents, or less formally, "real world documents", comprise of a mixture of images, text, signatures, tables, logos, water-marks, stamps, etc, and are often available only in scanned hardcopy formats. Search systems for such document collections are currently unavailable.

We describe our efforts at building a complex document information processing (CDIP) prototype. This prototype integrates "point solution" (mature) technologies, such as OCR capability, signature matching and handwritten word spotting techniques, search and mining approaches, among others, to yield a system capable of searching "real world documents". The described prototype demonstrates the adage that "the whole is greater than the sum of its parts".

To evaluate our CDIP prototype as well as to provide an evaluation platform for future CDIP systems, we also introduced a complex document benchmark. This benchmark is currently in use by the National Institute of Standards and Technology (NIST) Text REtrieval Conference (TREC) Legal Track. The details of our complex document benchmark are similarly presented.

Having described the global approach, we describe some additional point solutions developed in the IIT Information Retrieval Laboratory. These include an Arabic stemmer and a natural language source integration fabric called the Intranet Mediator. In terms of stemming, we developed and commercially licensed an Arabic stemmer and search system. Our approach was evaluated using benchmark Arabic collections and favorably compared against the state of the art. We also focused on source integration and ease of user interaction. By integrating structured and unstructured sources, we designed, implemented, and commercially licensed our mediator technology that provides a single, natural language interface to querying distributed sources. Rather than providing a set of links as possible answers, the described approach actually answers the posed question. Both the Arabic stemmer and the mediator efforts are likewise discussed.

A summary of the efforts discussed is found in [1].

Reference

1. Frieder, O.: On Searching in the 'Real World'. In: Argamon, S., Howard, N. (eds.) Computational Methods for Counterterrorism, ch. 1. Springer, Heidelberg (2009) ISBN: 978-3-642-01140-5

Y.A. Feldman, D. Kraft, and T. Kuflik (Eds.): NGITS 2009, LNCS 5831, p. 1, 2009.

Structured Data on the Web

Alon Y. Halevy

Google Inc.,
1600 Amphitheatre Parkway,
Mountain View, California, 94043,
USA
halevy@google.com

Abstract of Plenary Talk

Though search on the World-Wide Web has focused mostly on unstructured text, there is an increasing amount of structured data on the Web and growing interest in harnessing such data. I will describe several current projects at Google whose overall goal is to leverage structured data and better expose it to our users.

The first project is on crawling the *deep web*. The deep web refers to content that resides in databases behind forms, but is unreachable by search engines because there are no links to these pages. I will describe a system that *surfaces* pages from the deep web by guessing queries to submit to these forms, and entering the results into the Google index [1]. The pages that we generated using this system come from millions of forms, hundreds of domains and over 40 languages. Pages from the deep web are served in the top-10 results on google.com for over 1000 queries per second.

The second project considers the collection of HTML tables on the web. The WebTables Project [2] built a corpus of over 150 million tables from HTML tables on the Web. The WebTables System addresses the challenges of extracting these tables from the Web, and offers search over this collection of tables. The project also illustrates the potential of leveraging the collection of schemas of these tables.

Finally, I'll discuss current work on computing aspects of queries in order to better organize search results for exploratory queries.

Keywords: Deep web, structured data, heterogeneous databases, data integration.

References

1. Madhavan, J., Ko, D., Kot, L., Ganapathy, V., Rasmussen, A., Halevy, A.: Google's deep-web crawl. In: Proc. of VLDB, pp. 1241–1252 (2008)
2. Cafarella, M.J., Halevy, A., Zhang, Y., Wang, D.Z., Wu, E.: WebTables: Exploring the Power of Tables on the Web. In: VLDB (2008)

Y.A. Feldman, D. Kraft, and T. Kuflik (Eds.): NGITS 2009, LNCS 5831, p. 2, 2009.

Worldwide Accessibility to Yizkor Books

Rebecca Cathey[1], Jason Soo[2], Ophir Frieder[2], Michlean Amir[3], and Gideon Frieder[4]

[1] Advanced Information Technologies
BAE Systems
Arlington, VA
rebecca.cathey@baesystems.com
[2] Information Retrieval Laboratory
Illinois Institute of Technology
Chicago, IL
{soo,ophir}@ir.iit.edu
[3] Archives Section
United States Holocaust Memorial Museum
Washington, DC
mamir@ushmm.org
[4] School of Engineering and Applied Science
George Washington University
Washington, DC
gfrieder@gwu.edu

Abstract. Yizkor Books contain firsthand accounts of events that occurred before, during, and after the Holocaust. These books were published with parts in thirteen languages, across six continents, spanning a period of more than 60 years and are an important resource for research of Eastern European Jewish communities, Holocaust studies, and genealogical investigations. Numerous Yizkor Book collections span the globe. One of the largest collections of Yizkor Books is housed within the United States Holocaust Memorial Museum. Due to their rare and often fragile conditions, the original Yizkor Books are vastly underutilized. Ongoing efforts to digitize and reprint Yizkor Books increases the availability of the books, however, the capability to search information about their content is nonexistent. We established a centralized index for Yizkor Books and developed a detailed search interface accessible worldwide, capable of efficiently querying the data. Our interface offers unique features and provides novel approaches to foreign name and location search. Furthermore, we describe and demonstrate a rule set to assist searches based on inaccurate terms. This system is currently under the auspices of the United States Holocaust Memorial Museum.

1   Introduction

In contrast to traditional research efforts where the focus is on the design and evaluation of novel approaches or metrics, we describe an infrastructure oriented, research and development effort whose aim is to identify the location of and provide global access to information about Yizkor Books. We describe our system and the challenges we faced when creating such a system. This paper further describes the project initially outlined in [9].

Y.A. Feldman, D. Kraft, and T. Kuflik (Eds.): NGITS 2009, LNCS 5831, pp. 3–12, 2009.
© Springer-Verlag Berlin Heidelberg 2009

Specifically, we created a database of Yizkor Books metadata that enables search by multiple attributes such as location, including alternate spellings or common misspellings of the location, language, keywords, as well as text search of the abstract and table of contents. Links to digitized versions or translations are included when available. Through feature-rich search, the user is able to obtain greater insight into the contents of a book without actually seeing it. Furthermore, images such as maps and photographs may be associated with a book's entry. We broaden the relevance of Yizkor Books to more than genealogical research and research into a specific region's history by allowing search of the table of contents, abstracts, and keywords. This provides awareness of aspects shared by multiple locations that may not otherwise be detectable. Our database is available online and allows researchers from around the world to find potential wealths of information.

2   Overview of Yizkor Books

Yizkor Books memorialize life before, during, and after the Holocaust. They describe life in small and large towns as well as regions that may cover several countries. Yizkor Books were written by survivors of the Holocaust and others who came from the same town or area as a tribute to former homes, friends, and ways of life. They are commemorative volumes that contain documentation on destroyed communities and on the people who perished. They describe everyday life including some information on births, marriages, and deaths. In addition, they include anecdotes and character sketches, as well as information about local groups and organizations. Yizkor Books are feature rich, often containing detailed maps, photographs, illustrations, and necrologies (lists of those who perished).

Due to their diverse origin, Yizkor Books were published mainly in Yiddish, Hebrew, and English as well as parts in ten different languages starting in the early 1940s and continuing to the present where the highest activity occurred during the 1960s and 1970s [2]. Most Yizkor Books were printed in limited numbers by small organizations of survivors. Due to their limited number originally published, theft, and their fragile physical state, these original volumes reside in isolation and are under strict protection. Yizkor Books are a valuable resource for research into Eastern European Jewish history, Holocaust studies, and Jewish genealogy [6].

Several organizations have ongoing efforts to increase the availability of Yizkor Books for research. The New York Public Library has digitized 650 Yizkor Books and made them available to the public [1]. The books are listed alphabetically by place name. The Yiddish Book Center [2] has digitized the David and Sylvia Steiner Yizkor Book Collection containing over 600 Yizkor Books and made them available in new reprint editions [7]. JewishGen has a Yizkor Book Project [3] which seeks to facilitate access to Yizkor Books by providing links to digitized versions and available translations. If a translation is not available, they provide information to allow the user to contact a qualified translator. Both the Yiddish Book Center and the JewishGen Yizkor Book

---

[1] http://www.nypl.org/research/chss/jws/yizkorbookonline.cfm

[2] http://yiddishbookcenter.org/+10154

[3] http://www.jewishgen.org/yizkor

Project allow search for books by location. JewishGen also provides a necrology search of the lists contained in their books. Methods to find relevant books based on other criteria besides location or names are currently nonexistent. Furthermore, searching for books that contain information about an aspect of life shared across multiple locations is not possible. Although relevant books exist, they cannot be found due to variations of town names or the nonexistence of specified search features. Due to these facts, Yizkor Books are vastly underutilized. Our effort integrates Yizkor Book sources by developing a common metadata database and resolves many of the existing search limitations.

3   Challenges

Due to the diverse nature of Yizkor Books, several unique challenges were important to consider when designing a feature-rich Yizkor Book metadata database.

- **User Community:** Readers of Yizkor Books vary in nature. They may be knowledgeable, computer savvy research scholars or elderly individuals intimidated by computer technology. Their interests range from finding specific family members to completing dissertations on specific events. Some are Holocaust survivors themselves. It is critical to support all of these users. Thus, a diversity of search modes that meet the particular need of each type of user must be maintained.
- **Collection Language:** Yizkor Books contain accounts in thirteen different languages[4] – 62% Hebrew, 24% Yiddish and the other 11 languages making up the remaining 14% as the primary language. Almost every book was written using multiple languages. Users are not literate in all the languages; thus, the interface must, to the extent possible, support concise language independent interaction. Furthermore, the various spellings of names and locations is expected given their foreign language origin and must be accounted for.
- **Town Names:** Some town names are popular; consequently several towns in different areas of Eastern Europe have identical names. In some cases, town names have 15 variations. For example, what is now called Navahrudak at one time was known as Novogrudek or Nawardok [2]. The search architecture must ensure that users find the proper town even under these adverse conditions.
- **Efficiency:** In 1973, Yizkor Books already consisted of over 150,000 pages with portions written by more than 10,000 authors [10]. The collections of these books have grown a great deal since then. Simultaneous efficient search access for Web users must be guaranteed; yet resource constraints dictate that the host and back-up computers be strictly cheap commodity workstations.
- **Content:** Yizkor Books contain valuable information memorializing the victims of the Holocaust. While the most notable content is narratives from those involved in these tragedies, it also includes detailed town by town listings of the history of the Jews in a town from their first settlement there, their leaders over many years, education, community self help organizations, maps of the town with the Jewish quarter detailed, lists of professionals, lists of victims, the story of the Holocaust

---

[4] Hebrew, Yiddish, English, Czech, Dutch, French, German, Hungarian, Lithuanian, Polish, Romanian, Serbo Croatian, and Spanish.

in the town, as well as other aspects of life before, during, and after the Holocaust. In addition, Yizkor Books are feature rich, containing detailed maps, illustrations, photographs. When available, these items may be added to the database to provide more knowledge about the contents of a book.

4   Approaches

We used Ruby on Rails and MySQL to develop our system. Since users were assumed to be novice in nature, an AJAX based approach provided for selection guidance, namely interactive assistance based on the data stored in the server. Furthermore, using AJAX reduced the volume of data transferred since erroneous entries were not sent and likewise reduced the perceived user response time.

4.1   Collection Languages

Developing this system was difficult since users query the system for results in multiple languages without literacy in those languages. "On-the-fly" machine translators, if at all available considering the diversity of languages, produced inaccurate translations. Rather than present the user with garbled text, a set of parameters was designed (see Figure 1). These parameters allow users to select preferred categories to narrow their collection clusters.

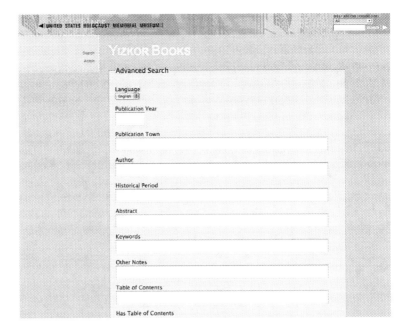

Fig. 1. Search Template

Our interface reduced query drift resulting from imprecise translations, providing the user with acceptable precision, where acceptable is based on informal user feedback. Even if the user is unable to read or write a language, they can select the language they are looking for, a publication year, a publication town, author, etc., from the selection fields, thus retrieving their intended results.

4.2   Town Names

Several towns have similar sounding names. Likewise, some towns have a diversity of spellings. Using predictive name expansion, via the AJAX techniques employed, users can select the correct town from the list of possible choices. Each keystroke that the user enters is searched through the database. As the keys progress, possible results that match the currently entered characters are shown below the search box. A direct search interface with automated spelling correction is available should the user prefer.

4.3   Efficiency

Efficiency is critical, even potentially at the expenses of some feature expansion. As such, for the relational search component, many of the relations are denormalized. Although arguable whether the denormalization of the tables is warranted, testing without such redesign indicated the need for faster (namely less join) processing. On average, query results are returned in less than 200ms when executed on a Dell Workstation (Intel Pentium 4 CPU 2.40GHz, 1GB RAM).

4.4   Text Search

Currently, we support only plain-text and rudimentary search capability. The integration of XML tagged document search and complete retrieval relevance ranking capability using relational technology is under development. That is, we are integrating SIRE[5] [5] with the Yizkor Book Collection. SIRE provides full information retrieval functionality in a strictly relational platform. Such functionality includes leading similarity measures, proximity searching, n-grams, passages, phrase indexing, and relevance feedback.

Similarly, we plan to likewise incorporate our SQLGenerator [4]. The SQLGenerator supports a complete suite of XML search features, and like SIRE, executes on a purely relational platform. Since the Yizkor Book Collection is generally available on relational platforms, both the SIRE and the SQLGenerator are ideal supplements to our existing capabilities.

5   Architecture

An efficiency challenge was the large number of one to many relationships needed to model Yizkor Book information. Since there are many variations of spellings for the title, a book can have many different titles. Similarly, a book may contain information

---

[5] Scalable Information Retrieval Engine.

about multiple towns, regions, or countries. A contributor is defined as an author, editor, or translator. Multiple people contribute to a book entry. An entry can also contain images such as maps, photographs, or detailed drawings. Likewise, a book can be written in multiple languages and contain information about a variety of subjects.

Our relational architecture is designed around a main books table. The books table links to tables giving information about locations, contributors, subjects, and images. Since a book can have alternate titles, the books table also links to an alternate titles table. Similarly, a location can have variations. Therefore, the locations table also links to a location variation table.

6   Features

The search interface embeds various features including but not limited to: automated query completion, ranked corrected term suggestions, and the restoration of deleted records. These features benefit both researchers and the administrative staff.

6.1   Restoration Feature

To simplify the operational management of the system, a restoration feature was developed. This feature allows the staff to view deleted records from the past and restore them. There are two types of restorations: range and selection. Range restoration restores all deleted records within a given date of deletion range, whereas selection restorations are for specific records.

6.2   Spell Checking

As previously discussed, Yizkor Books are written in a diversity of languages. Thus, regardless of the language proficiency of the user, many of the terms within the Yizkor Books are foreign to her/him. Furthermore, many geographical locations and personal names are spelled similarly, particularly to a user that is unfamiliar with their origin. We now describe our simplistic, yet statistically successful, endeavor to assist users search in this multi-language, similarly sounding and spelled entity environment.

Based on our experience with foreign language stemming [1] and foreign name search [3] we know that simplicity is of the essence; hence, we crafted six simplistic rules to nullify the effects of misspellings. As shown in our prior efforts, such simplistic rules are relatively efficient and tend to perform well. Our rules are shown in Figure 2 and explained in detail in [9].

Initially, we issue the query. If a result is obtained, we assume this result to be valid. Should no results be obtained, another query is issued automatically, this time with a set of modified strings. For each result now returned, we compute its similarity to the original query and determine a confidence with respect to the desired name. The global confidence of the merged result set is computed. If the global confidence fails to meet the needed threshold, an n-gram based solution [8] is deployed. The global rule based confidence and the n-gram generated confidence are compared, and the spelling variants generated by the approach with the higher confidence are suggested to the user. The threshold is determined based on the level of certainty of the user that s/he spelled the term correctly or the assumed default level of certainty of the user when such information is unknown. The greater the user level of certainty, the lower is the tolerance of misspelling, namely the lower is the degree of modification.

1. Replace first and last characters from query string with SQL search wild cards (%). Repeat string substitution until a sufficient confidence in the results returned is obtained or no additional results are found.
2. Replace middle character with SQL search wild card. Repeat string substitution.
3. Replace first half of query with SQL search wild card.
4. Replace second half of query with SQL search wild card.
5. Retain first and last characters, insert SQL search wild card in between retained characters.
6. Retain first and last 2 characters, insert SQL search wild card in between retained characters.

Fig. 2. Rule Set to Identify Misspellings

The alternate spelling generation rules potentially generate an extensive set of candidates especially in the case of lengthy queries. To prevent long execution times, we imposed execution termination criteria on each of the rules. These include limiting the number of potential candidates (currently set at 20), limiting the retrieval time from the database (currently set at 150 ms), and constraining the number of repetitions in each repetitive rule.

Assume the desired query was: "Z'eludok". If the query was correctly spelled, the results presented in Figure 3 would be displayed to the user. Assume, however, that

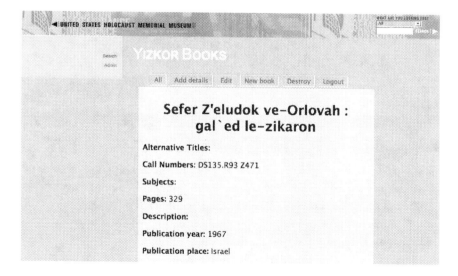

Fig. 3. Results

Table 1. Example Phases. Query: Zeludoc.

Algorithm	Phase 1	Phase 2	Phase 3
First	%eludo%	%lud%	%u%
Second	Zel%doc	Zel%oc	Ze%oc
Third	%udoc	-	-
Fourth	Zel%	-	-
Fifth	Z%c	-	-
Sixth	Ze%oc	-	-

Fig. 4. Rankings

Table 2. Correct Candidate Ranking Over 100 Random Queries

Rank	Percentage
First	73%
Second	9%
Not Listed	18%

the user incorrectly typed the title "Zeludoc." In such a case, our approach, after not retrieving any entries, would attempt to correct the spelling. Such correction using our rule set would derive the candidates illustrated in Table 1 and would present the ranked candidates to the user as shown in Figure 4. Our preliminary tests show the correct result typically appeared in the top two results, for the top 100 random queries used (see Table 2). Frequently, the correct result appeared as the top candidate.

7 Accuracy Results

We compared the accuracy of our integrated six rule approach against a traditional n-gram based solution. We used a randomly selected subset of roughly 1000 names from

Table 3. Percentage of Correct Names Found

	INSERT			DELETE			REPLACE			INVERT		
# chars	2	3	4	2	3	4	2	3	4	2	3	4
n-gram	95.45	91.61	86.85	88.62	79.95	75.58	85.04	71.27	61.24	57.69	48.08	40.17
rules	99.70	98.39	95.87	99.87	93.81	89.00	94.74	86.62	77.49	72.85	61.37	51.95

the 1942 Slovak Jewish Census. Names averaged 12 characters in length, with a median length of 8, and a standard deviation of 3.61 characters. We randomly added, removed, replaced, and inverted characters in random locations. Deletions were limited so that terms remained at least 4 characters.

Table 3 shows the percentage of correct names found using a traditional n-gram approach and our integrated n-gram and rule set approach. The columns represent insertion, deletion, replacement, and inversion of 2 to 4 characters. The position of character(s) in the string was randomly generated using a uniform distribution. In the case of multiple character inversion, randomly chosen pairs of characters were exchanged sequentially. In all configurations, our integrated n-gram rule set based approach correctly identifies a higher percentage of names than a traditional n-gram solution.

8   Ongoing Modifications

Our efforts act as a mediator between the user and bibliographic data of the Yizkor Books, providing users with a uniform interface to search the collection. Shortly, via other available means, users will be able to browse and search the exact pages, cover to cover. Most of these books are already available in digitized format provided by the New York Public Library[6].

An additional goal of this effort is to allow for user commentary of the books within the collection. This commentary can be of value in several different ways. One primary way is to allow those who were involved in these events to quickly and easily add their unique accounts. These accounts will then be stored and will further increase the value of the collection.

Furthermore, this annotation capability could increase collaborative research. At the end of a lengthy process, researchers could directly publish any findings they may have and permanently attach it to the book(s). This would allow future researchers to instantly view, assess, and make use of past findings.

9   Conclusion

We designed, developed, and deployed an integrated structured and unstructured search system for the bibliographic information on the Yizkor Book Collection, a collection that is of significant historical and humanitarian interest. Our site provides a valuable, and previously nonexistent, service to researchers by providing a specified search interface.

---

[6] http://yizkor.nypl.org/

Extensive testing demonstrated an average query response time of just under 200ms. This system is online and publicly available. Detailed information about the books are currently being added. Prior to this effort, no centralized, extensive, in depth bibliography existed for this collection nor any other collection of its nature.

References

1. Aljlayl, M., Frieder, O.: On Arabic Search: Improving the Retrieval Effectiveness via a Light Stemmer Approach. In: Proceedings of the 11th ACM Conference on Information and Knowledge Management (CIKM 2002), Washington, DC (November 2002)
2. Amir, M.: From Memorials to Invaluable Historical Documentation: Using Yizkor Books as Resource for Studying a Vanished World. In: Annual Convention of the Association of Jewish Libraries, vol. 36 (2001)
3. Aqeel, S., Beitzel, S., Jensen, E., Grossman, D., Frieder, O.: On the Development of Name Search Techniques for Arabic. Journal of the American Society for Information Science and Technology (JASIST) 57(6) (April 2006)
4. Cathey, R., Jensen, E., Beitzel, S., Grossman, D., Frieder, O.: Using a Relational Database for Scalable XML Search. Journal of Supercomputing 44(2) (May 2008)
5. Frieder, O., Chowdhury, A., Grossman, D., McCabe, M.: On the Integration of Structured Data and Text: A Review of the SIRE Architecture. In: DELOS Workshop on Information Seeking, Searching, and Querying in Digital Libraries, Zurich, Switzerland (December 2000)
6. Kasow, H.: Developing a Jewish Genealogy Library: The Israel Genealogical Society Library as a Case Study. In: Proceedings of IFLA Council and General Conference (August 2000)
7. Madsen, C.: In the Bond of Life. Pakn Treger (Summer 2005)
8. Manning, C., Raghavan, P., Schutze, H.: Introduction to Information Retrieval. Cambridge University Press, Cambridge (2008)
9. Soo, J., Cathey, R., Frieder, O., Amir, M., Frieder, G.: Yizkor Books: A Voice for the Silent Past. In: Proceedings of the 17th ACM Conference on Information and Knowledge Management (CIKM 2008), Napa, CA (October 2008)
10. Wein, A.: Memorial Books as a Source for Research into the History of Jewish Communities in Europe. In: Yad Vashem Studies on the Eastern European Catastrophe and Resistance, vol. IX (1973)

Biomedical Information Integration Middleware for Clinical Genomics

Simona Rabinovici-Cohen

IBM Haifa Research Lab
Haifa University, Mount Carmel, Haifa 31905, Israel
simona@il.ibm.com

Abstract. Clinical genomics, the marriage of clinical information and knowledge about the human or pathogen genome, holds enormous promise for the healthcare and life sciences domain. Based on a more in-depth understanding of the human and pathogen molecular interaction, clinical genomics can be used to discover new targeted drugs and provide personalized therapies with fewer side effects, at reduced costs, and with higher efficacy. A key enabler of clinical genomics is a sound standards-based biomedical information integration middleware. This middleware must be able to de-identify, integrate and correlate clinical, clinical trials, genomic and images metadata from the various systems. We describe MedII, a novel biomedical information integration research technology that some of its components were integrated in IBM Clinical Genomics solution. We also introduce the need for biomedical information preservation to assist in ensuring that the integrated biomedical information can be read and interpreted decades from now.

Keywords: information integration, biomedical systems, de-identification, long term digital preservation.

1 Introduction

On June 26, 2000, the first draft of the human genome was announced, significantly boosting research on correlations between phenotypic and genotypic data. Today, some genomic data is produced by DNA microarrays that generate high throughput molecular data and by several hundred non-expensive genetic tests that probe small portions of the genome. While sequencing a full human genome is still very costly, taking approximately six months and $10 million to $50 million to complete, new cheaper and faster sequencing machines are being developed (such as the Helicos BioSciences machines). It is assumed that within a few years, a human genome could be completely sequenced in one day at the cost of $1000, creating new healthcare and life sciences practices in which sequencing an individual's genome becomes a commodity (see the NIH request for application number RFA-HG-08-008 titled "Revolutionary Genome Sequencing Technologies – The $1000 Genome").

Consequently, a new field known as Clinical Genomics is emerging in the healthcare and life sciences domain that may contribute to a revolution in the health of humankind.

Y.A. Feldman, D. Kraft, and T. Kuflik (Eds.): NGITS 2009, LNCS 5831, pp. 13–25, 2009.
© Springer-Verlag Berlin Heidelberg 2009

Clinical genomics is the correlation of clinical information—such as patient records including environmental data, family histories, medications, and lab tests—with knowledge about the human or pathogen genome. This correlated information will impact decisions about diagnosis, prognosis, treatment, and epidemiology. By understanding illnesses on the molecular level, including gene variations linked to disease or drug response, doctors may be able to make more precise diagnoses and tailored treatment decisions. Pharmaceuticals researchers will discover new drugs and develop targeted treatments that have better safety and higher efficacy. Clinical genomics will also improve healthcare guidelines and protocols, leading the way towards truly personalized healthcare and information-based medicine.

However, realizing clinical genomics has difficulties. Much of the data is still in propriety formats in various silos, and privacy issues associated with sharing that data still exist. A biomedical information integration technology is needed that can standardize, de-identify, integrate, and correlate clinical trials with genomic and image metadata described in diverse formats and vocabularies and scattered in disparate islands of data. The data needs also be integrated with public data sources such as PubMed and GenBank.

We describe a novel research technology, named MedII (previously known as Shaman-IMR), for biomedical information integration, which some of its components were incorporated in IBM Clinical Genomics solution [1]. MedII includes a set of services to standardize, de-identify, integrate and correlate the various biomedical data. The services utilize XML technology [2], Model Driven Development (MDD), and emerging healthcare and life science standards. XML technology is well-situated to serve as the glue for integrating the various data sources because XML includes both data and metadata in a universal format, namely text, and it is agnostic to the platform or application used to create or consume the data. However, XML lacks semantic at its core, so standards are needed to provide the semantics, and indeed vast standardization efforts by various organizations are witnessed. MDD aims to separate domain knowledge from the underlying technology specifics, enabling to build tools that ease the generation and update of transformation services and data models.

MedII utilizes an integrated target schema that is a union of sub-schemas; one for each standard in the domain. The records of the various standards are correlated via a global privacy id which is the same for all documents of the same individual, as well as via standardized controlled codes and vocabularies. This MedII approach reduces cost and complexity as a new data source can be added or removed without changes in the other data sources or target schema. Moreover, data de-identification is an integral part of the system and the generated privacy id is used to replace all the identifying information as well as to correlate among the records of the same individual.

The rest of this paper is organized as follows. The next section provides related work and our summarized contribution. Then, we describe an overview of MedII various layers followed by additional three sections that dive deeper into those layers. The integration layer section describes the services, which generally run in the source premises, to standardize and enrich the data so that it could be integrated with data from other sources. The de-identification section describes a service which is required by government legislation in the lack of user consent, and thus has special interest. The index layer section describes the services to manage and index the data in order to be efficiently queried and searched by various applications. Finally, as much of the integrated data should be preserved for decades, we describe the need for biomedical information preservation and propose future work in that space.

2 Related Work

The direct effect of healthcare and life sciences on each individual, the ever growing volumes of biomedical data and the advances in genomics encouraged research in biomedical information integration for several years [3]. One approach, also termed "integration at the glass", performs integration in the eyes of the human user by putting data from different sources side by side on the same screen. PharmGKB [4] which integrates literature, diseases, phenotypes, drugs and genomic information is an example of such approach. This approach is sometimes combined with data grids such as in the BIRN [5] project funded by NIH, which is a data grid that targets shared access to neuro-imaging studies. While this approach is effective for human users, it is not adequate for machines which need to process and mine the data.

Other approaches offer integration which either merges the query result sets (e.g. OPM [6], DiscoveryLink [7]), or merges the sources data (e.g. GenMapper [8], Bio-Mediator [9]) and thus are adequate for human and machine users. OPM offers a propriety query language to join remote databases via wrappers. DiscoveryLink which is now embedded in IBM DB2 is a federation technology in which integration is done in query time using wrappers to the source databases. GenMapper integrated data from more than forty data sources into a 4-table propriety schema. BioMediator is similar but uses a more elaborated target schema defined using the Protégé knowledgebase system. All these projects are schema focused and tend to employ propriety schemas, and require learning propriety and complicated languages. Creating the semantic mappings from the sources to the target schema is tedious and is generally done manually, but this may be relaxed by using automatic schema mapping methods [10]. Also, these projects do not deal with privacy issues associated with the data.

Aladin [11] is a system which combines the schema focused approach with a data focused approach and utilizes automatic mechanisms for finding links among source objects. Aladin is mostly for biological data and it assumes the data is semi structured and text centric rather than a structured data centric database. Adding or removing data sources as well as updates to the current data sources are expensive in the Aladin system because all links must be recomputed even if only a small fraction of the data of a data source changed.

In MedII, each data source is cleansed, augmented with terminology codes and transformed to the standard for that data type, in case the source data wasn't in the required standard format initially. Then, the standardized data is assigned with a privacy id, de-identified and stored in a data model based on that standard. Thus, the integrated schema is a union of sub-schemas; one for each standard. The records of the various standards are correlated via the global privacy id which is the same for all documents of the same individual, and via the standardized controlled codes and vocabularies.

MedII offers integration which is adequate for a human or machine user. It employs standard-based schemas and does not require learning any propriety language. Using MDD based tools, data transformation and generation of data models is simplified. Additionally, adding, removing, or updating a data source has a contained impact and there is no effect on the other data sources or the target schema. Finally, in MedII, data de-identification is an integral part of the system and the generated privacy id is used to replace all the identifying information as well as to correlate among the records of the same individual.

3 MedII Conceptual Layers Overview

Clinical and genomic data are scattered among different archives (computerized and paper-based) in various locations. Data is generally stored where it was created, and is not always available to researchers. Furthermore, this information is often expressed using different vocabularies, terminologies, formats, and languages, and is retrieved using different access methods and delivery vehicles. Some of the data resides in external open databases, which include rich data but are not easy to navigate.

MedII goal is to integrate those diverse data sources with a low cost and while considering privacy limitations. The cost remains minimal even when adding, removing or updating a data source. Update in one source does not cause updates to the integration of the other data sources. Similarly, adding or removing a data source does not impact the other parts of the system.

MedII uses standardized formats and workflows which facilitates the transition from information in silos to cross-institutional information integration. It includes three conceptual layers as shown in Figure 1 below. The first is an **integration layer**, which includes tools and services for transforming data from propriety formats to standard XML-based representation. This layer also includes data cleansing and enrichment services, such as adding codes from controlled vocabularies or adding annotations from public data sources. Finally, it includes a de-identification service for de-identifying protected data prior to leaving its source premises, as required by government legislation. The **index layer** generates efficient standardized data models and indexes for structured, semi-structured and unstructured data so that applications on top can query, perform free text search and retrieve the data in an efficient and powerful way. The **EHR layer** (Electronic Health Record layer) is designed for applications in the healthcare domain, and creates longitudinal and cross-institutional individual-centric objects that are compiled from the documents indexed by the previous layer. As this layer is not an integral part for the integration, it is not discussed further in a separate section. Mining and query tools, predictive engines or other applications can reside on top of either the second layer or the third layer [12]. MedII facilitates complex queries such as: "What protocols were used for tumors that produced similar staining sections and were from patients aged 40-60 with the same 'Yakamura' polymorphism in their genes?" The various services of MedII layers can be exploited in a Service Oriented Architecture and attached to an enterprise service bus.

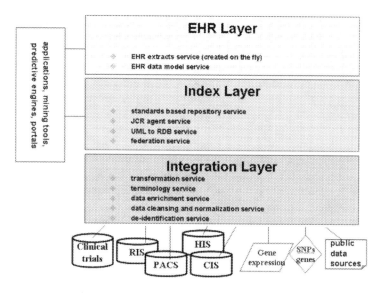

Fig. 1. MedII layers and services

4 Integration Layer

The integration layer includes several services, only some of which may be used and the order in which those services are called is flexible. The services employ MDD and include a transformation service, a terminology service, a data enrichment service, a data cleansing and normalization service, and a de-identification service, required by law in the lack of user consent. The de-identification service is described in more detail separately in the next section. Usually, it is preferable to apply the transformation service first because then you can employ the other services on the obtained standardized formats, thereby simplifying their employment.

The transformation service transforms the various proprietary biomedical data formats to standardized formats, or transforms one standard to another, thereby enabling the generic correlation of data from cross-institutional sources while reusing the domain knowledge exploited by the standard organizations. This knowledge is obtained over many years, from various contributors coming from diverse organizations. In some cases, only some of the data to be shared is transformed to a standardized format along with a link to the full original data. These cases are usually known as metadata extraction and occur primarily when some of the data is in an uninterpreted format, such as binary format. For example, in DICOM images, only some of the DICOM tags are extracted and transformed to a standard along with a link to the full DICOM object in the original Picture Archive Communication Systems (PACS) system.

Biomedical data includes various types of information, including clinical data from Clinical Information Systems (CIS) in hospitals and HMOs, demographic data from Hospital Information Systems (HIS), diagnostic and imaging data from Radiology

Information System (RIS) and PACS, clinical trials data, environmental and life style data, and genomic data that partially reside in Laboratory Information System (LIMS). Currently, the standards for representing biomedical data are still emerging and some of them have not yet been widely adopted. Some data types do not have standards, while other data types have several standards. Most of the new standards are XML-based and still evolving. As a result, one of the challenges is to decide what standard to use for what data.

The standards for clinical data include the widely used HL7 V2 and its XML version HL7 V2.XML. HL7 is also developing the new V3 family using MDD methodology, including among others the Clinical Document Architecture (CDA) and Clinical-Genomics message, but this new family is still in the early adoption phase. Digital imaging data is dominated by the DICOM standard. The new DICOM Structured Reports (SR) for imaging reports is still in early adoption phases.

The standards for genomic data include OMG MGED MAGE for gene expression and the related Minimum Information About a Microarray Experiment (MIAME), NIH HapMap for haplotype data, BSML for bioinformatic sequence, SBML for Systems Biology, and NCI caBIO for cancer bioinformatics infrastructure objects. All these standards are developed with MDD methodology and are still in the early adoption phase.

The standards for clinical trials data are CDISC ODM for operational data that is moved from a collection system to the central database of the sponsor of the clinical trial. CDISC SDTM is used to submit data to regulatory authorities such as the FDA. CDISC Define.xml is an extension of ODM that includes metadata on the domains and variables used in SDTM. All these standards are still in early adoption phases.

In many standards organizations and large consortiums, domain knowledge is increasingly captured by domain experts using UML conceptual models, which include data elements without methods. In the MDD approach, the source and target schemas are represented in UMLs, and UML profiles are introduced to extend and constrain UMLs for a particular use. Accordingly, the transformation service, which utilizes the MDD approach, includes a transformation tool that imports the source and target UML models and then configures the UML profile and mappings from the source to the target schema. By using MDD, the service becomes generic and agnostic to changes in the models that occur during the standardization process.

The terminology service augments the data with codes from controlled vocabularies, thus facilitating more accurate search and mining of the integrated data. The primary controlled vocabulary is SNOMED, a very extensive vocabulary adapted by the US government and available free for distribution. Additional controlled vocabularies include LOINC for laboratory results, test orders and document types; ICD-9 and ICD-10 for diseases; RxNORM, which is adapted by FDA for medications; CPT for procedures and genomic tests; and HGNC for nomenclature of gene symbols and names. The terminology service can be built on top of open existing terminology services such as HL7 Clinical Terminology Service (CTS) or NCI Enterprise Vocabulary Services (caEVS).

The data enrichment service adds to the data annotations and related data from public data sources. PubMed is a prominent public source for clinical data that includes medical publications. Additionally, there are rich public data sources for genomic data, such as EBI ArrayExpress for microarray data, GenBank for genetic sequence, dbSNP for simple genetic polymorphisms, UniProt and PDB for proteins, and OMIM for correlations of human genes and genetic disorders. The relevant content of those public data sources are sometimes embedded within the standardized data to be shared, while in other cases it is federated in the index layer. The first case is used when it is important to document the evidence that brought to a specific conclusion, e.g., for compliance purposes. The latter case is preferred when it is important to enrich the data in real time with the most recent information from public sources.

In MedII, we implemented transformations to HL7 V3, specifically CDA, as well as to MAGE, ODM, SDTM, BSML, and performed metadata extraction from DICOM. We have used the SNOMED, LOINC and ICD9 terminologies and performed a federation with PubMed public data source. However, there are no limitations to adding additional standards, terminologies or public data sources to the system.

5 De-identification

De-identification issues usually arise when enterprises need to share data that is related to individuals or includes sensitive business data. For example, in clinical genomics, de-identification is often mandated when cross-institutional data needs to be integrated and correlated. Medical information is naturally associated with a specific individual, and when this data leaves the source premises, it must be altered so that it cannot be re-associated with that individual by the recipient. Even in this case, the process should preserve the ability to correlate the de-identified document with other de-identified documents or records from the same individual.

Government guidelines and legislations in various countries define Protected Health Information (PHI) data, which if lacking user consent must be de-identified before leaving the data source premises. This includes the Health Insurance Portability and Accountability Act of 1996 (HIPAA) in the United States, the Freedom of Information and Protection of Privacy Act (FIPPA) in Canada, and the Personal Information Protection and Electronic Document Act (PIPEDA) in the European Union, Australia, and New Zealand. For genomic data, there is also the NIH-DOE Joint Working Group on Ethical, Legal, and Social Implications of Human Genome Research (ELSI).

HIPAA, the least stringent of the legislation listed above, defines seventeen identifiers that must be de-identified. Some of the identifiers that must be removed includes the name, certificate/license numbers, diagnostic device ID and serial number, biometric identifier (e.g., voice, finger print, iris, retina), full face photo or comparable image, SSN, fax numbers, electronic mail address, URL, IP address, medical record number, health plan number, account numbers, vehicle ID, serial number, and license plate number. The other identifiers are less restrictive. In address identifiers, the city/town, state, and first three digits of zip code can be kept if the population in the city is greater than 20,000. Similarly, in date identifiers (e.g., DoB, ADT, DoD), the year can be kept, and in telephone numbers identifiers, the area code and prefix can be kept if geographical information is missing. Additionally, the eighteenth item defined by HIPAA is called "Other", and it refers to information that exists in free text sections of patient records and can be used to identify the patient. Examples of phrases that must be de-identified according to the "Other" identifier are "the British Queen" and "The patient suffered from serious leopard bite acquired at the De Moines Zoo fire incident". Note that according to HIPAA, gender, race, ethnic origin, and marital status can be kept.

In MedII, the UDiP technology is used within the integration layer to provide de-identification service, as depicted in Figure 2 below. UDiP includes a highly configurable and flexible engine, generally located on the source premises, that receives various types of data, including clinical, imaging, and genomic data, and various formats, including comma-separated-values (CSV) files, DICOM images, and XML data. UDiP de-identifies the PHI while maintaining the correlation between the various records belonging to the same individual, and then transfers it (pass the red line) to the shared data area. By transferring de-identified data to the shared area, we prevent access to the data in the source premises, thus avoiding the risk of a direct privacy attack on the original data. Maintaining the correlation between the various records is obtained by either PHI encryption or by using an Anonymous Global Privacy Identifier (AGPI) server. Although PHI encryption is a simpler method, it is not always preferable. According to some interpretation of HIPAA, PHI encryption violates the "Other" field by keeping the PHI (although encrypted) together with the medical data in the shared area.

With the AGPI server method, the PHI is de-identified (removed) and replaced by an AGPI opaque value assigned by an AGPI server. That AGPI server resides in a secured area with strict access control. Only authorized people can access it with an AGPI value for re-identification. Note that in this case, there is no focal point to include the PHI and the actual medical data – they are disjoint. The AGPI is used to maintain the correlation between documents belonging to the same individual, without identifying the individual. The AGPI server is required to fulfil two criteria. The server gives the same AGPI to the same individual (regardless of the source of the document, the spelling of the patient's name, incompatible fields, etc.), and it never gives the same AGPI to two different individuals.

Since privacy restrictions change with time and are distinctive in different locations, UDiP is highly configurable and extensible. A model-driven tool that can be used by domain experts configures the data types to de-identify, the locations of the PHI, the action to apply on each location, and the method for generating and storing the AGPI.

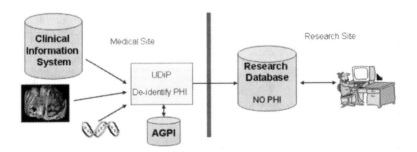

Fig. 2. De-identification with AGPI server

6 Index Layer

The index layer is responsible to manage and index all the normalized standardized documents obtained from the integration layer, so that applications on top can efficiently perform powerful queries, searches and data retrieval. It mainly contains information management services such as standard based repositories services and services that employ MDD to generate those repositories. The data models used in the index layer are standard based and include a separate schema for each standard. The records of the various standards are correlated via the AGPI which is the same for all documents of the same individual and via the standardized controlled codes and vocabularies. This design allows the flexible update of a data model for a specific standard that evolved without affecting the other data models of the other standards.

There are several options to manage and index the data that came from numerous sources. One option is the aggregation model in which the data is physically stored outside the sources premises in a shared repository and the indexation and queries are performed on that repository. In some cases, the repository can include some metadata of some other documents such as in the case of imaging where only the image metadata is kept in the repository with a link to the full image in the PACS system. Another option is the federation model in which no data is stored outside the sources premises. Instead, mappings are defined from the shared system to the sources. In query time, the query is divided to sub-queries which are posed against the various data sources. Then, the correspondent sub-results are jointed to one result to form the query result. In the federation model there is no data stored outside of the sources premises, but on the other hand it requires that the sources are on-line all the time and the performance of the queries is decreased with respect to the aggregation model.

A third option, which is used in MedII, is a hybrid model of aggregation and federation. Data from operational systems are aggregated to prevent them from interfering during operational time. Data from public sources that are not under institutional control are generally federated. Accessing the external reference data via federation ensures data currency, but at the cost of performance degradation and at the risk of security compromise.

The biomedical data includes structured as well as unstructured data. A case report form, for example, includes patient name, gender, age, and so forth, which is structured data, along with a description of the medical history, which is unstructured data. To query and search the data, there is a need for a combined relational database index and an information retrieval (IR) index. Today, the data is either treated as structured data and stored in a relational database, or treated as unstructured data and stored in a content management system. In the first case, the data is shredded (decomposed) in a relational database but then the free text search capabilities are limited. To this end, an IR technology is added to convert unstructured data to semi-structured content using some annotators, and then the associated meta-data can be joined with the relational repositories.

In the latter case where the documents are stored in a content management system, the queries for these documents are generally less powerful. For example, it's difficult to find patients that have two medications, where each medication appears in a different document (a.k.a. the join operation in the relational database). The new Java Content Repository (JCR) specification holds the promise to bridge the gap between the structured and unstructured worlds, and provide an interface to combine query and free text search. The JCR has a hierarchical data model and allows to store various structured and unstructured typed properties (leafs of the hierarchy) such as strings, binaries, dates. JCR query manager facilitates structured SQL queries as well as XPATH queries with free text search capabilities.

As stated earlier, the domain knowledge is increasingly represented in conceptual UML models. Standard organizations and government sponsored initiatives build public UML models for biomedical data representation as well as models for disease specific data. Examples of such models are HL7 RIM, OMG MAGE, NCI caBIO, and HUPO PSI. Moreover, it is assumed that pharmaceutical companies and academic medical research centers will also build conceptual UML models for their specific use cases by either tailoring those publicly available models or by building their own models.

Thus, the MDD approach is vastly used in the index layer as well. It includes tools and runtimes to semi-automatically generate a relational data model or a JCR data model for a given UML. The services such as the JCR agent service introduce UML profiles to extend and constrain the UMLs to describe primary keys, JCR nodes references, etc. According to the UML profile configuration, the JCR agent builds a JCR data model and the runtime receives the XML documents that are compliant with the defined UML model as input and stores them in the JCR-compliant repository. The UML to RDB service provides similar functionality but for the relational database case. Another example of an MDD based service is a service that builds a domain text annotator for a given model.

MedII has implemented the JCR agent and the UML to RDB services which generate standard based repositories. Additionally, it uses DB2 underneath, so it inherits the federation service of DB2.

7   Future Work – Biomedical Information Preservation

The amount of biomedical information is constantly growing and while some of the data includes small text objects like CDA documents, the data generated by medical devices include large (gigabytes) born-digital binary objects such as large images and gene expressions. To assure a lifetime EHR as is proposed nowadays by a new US bill, and also to support compliance legislations, this data should be preserved for the individual whole life time and even beyond that for research purposes and treatments of descendents.

This poses a new challenge to technologists – the challenge of biomedical information preservation, namely how to ensure that the biomedical data can be read and interpreted many years (tens or hundred years) from now when current technologies for computer hardware, operating systems, data management products and applications may no longer exist. As the cost of biomedical information integration is high, it is important to add to it digital preservation capabilities from the beginning, so that the integrated data will be interpretable for many years to come and the high cost is shared with the future over a long period.

A core standard for digital preservation systems is the Open Archival Information System (ISO 14721:2003 OAIS) [13], which targets the preservation of knowledge rather than the preservation of bits, and provides a set of concepts and reference model to preserve digital assets for a designated community. OAIS defines the preservation object that is the basic unit to be stored for preservation. It is the Archival Information

Package (AIP) which consists of the preserved information, called the content information, accompanied by a complete set of metadata.

Figure 3 below depicts an example AIP for a CDA document. The content data object is the raw data intended for preservation namely the CDA document itself. The representation information consists of the metadata that is required to render and interpret the object intelligible to its designated community. This might include information regarding the hardware and software environment needed to render the CDA or the specification of the CDA data format. The other AIP metadata, called the Preservation Description Information (PDI) is broken down by OAIS into four well-defined sections:

- **Reference information** - a unique and persistent identifier of the content information both within and outside the OAIS such as the CDA ClinicalDocument.id object.
- **Provenance information** - the history and origin of the archived object such as a description of the organization in which the CDA was created (ClinicalDocument.Custodian object).
- **Context information** - the relationship to other objects such as related X-rays, related lab tests, previous encounters for the same theme, consent document, encompassing encounter, etc.
- **Fixity information** - a demonstration of authenticity, such as checksums and cryptographic hashes, digital signatures and watermarks.

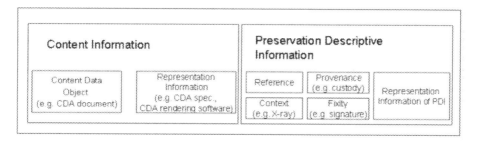

Fig. 3. Example AIP for CDA

At the heart of any solution to the preservation problem, there is a storage component, termed archival storage entity in OAIS. We argue that biomedical information integration systems will be more robust and have less probability for data corruption or loss if their storage component is a preservation aware storage, namely if their storage has built-in support for preservation. As more and more storage systems offload advanced functionality and structure-awareness to the storage layer, we propose to offload OAIS-based preservation functionality to the storage system. To that end we are developing Preservation DataStores (PDS) which realizes a new storage paradigm based on OAIS [14].

For future work, we propose building preservation objects for biomedical information and storing those preservation objects in an OAIS-based preservation aware storage such as PDS.

8 Conclusions

The recently new field of clinical genomics holds the promise to a more in-depth understanding of the human and pathogen molecular interaction. This will lead to a personalized medicine with more precise diagnoses and tailored treatment decisions. A key enabler of these new field is a biomedical information integration technology that can standardize, integrate and correlate the disperse data sources. Various clinical and genomic data sources are extremely valuable independently, but may contain even more valuable information when properly combined.

In this paper we described MedII, a biomedical information integration research technology that utilizes XML technology, model driven development, and emerging healthcare and life sciences standards. The use of standards and XML technology enable semantic integration in a universal format while using MDD makes the technology agnostic to standards evolvement and adaptive to each specific solution extension. MedII features flexible integration of new data sources or update of existing ones without affecting the other parts of the system. Furthermore, the various MedII services, that some of them are incorporated in IBM Clinical Genomics solutions, can be utilized within SOA architecture.

New technology advances and legislations will require preserving the individual integrated biomedical information for his whole lifetime and even beyond that for research purposes and treatments of descendents. We propose for future work to build preservation objects for the integrated biomedical information and utilize an OAIS-based preservation-aware storage for long-lived storage and access to those objects.

Acknowledgements. We would like to thank Barry Robson and Pnina Vortman who were visionary and initiated the Shaman-MedII project. Additional thanks to Houtan Aghili, OK Baek and Haim Nelken who managed IBM Clinical Genomics solution. Finally, thanks to the great team that helped creating MedII various components including Flora Gilboa, Alex Melament, Yossi Mesika, Yardena Peres, Roni Ram and Amnon Shabo.

References

1. IBM Clinical Genomics solution,
   http://publib.boulder.ibm.com/infocenter/eserver/v1r2/
   index.jsp?topic=/ddqb/eicavcg.htm
2. Shabo, A., Rabinovici-Cohen, S., Vortman, P.: Revolutionary impact of XML on biomedical information interoperability. IBM Systems Journal 45(2), 361–373 (2006)
3. Altman, R.B., Klein, T.E.: Challenges for biomedical informatics and pharmacogenomics. Annual Review of Pharmacology and Toxicology 42, 113–133 (2002)
4. Hewett, M., Oliver, D.E., Rubin, D.L., Easton, K.L., Stuart, J.M., Altman, R.B., Klein, T.E.: PharmGKB: the Pharmacogenetics Knowledge Base. Nucleic Acids Research (NAR) 30(1), 163–165 (2002)
5. Astakhov, V., Gupta, A., Santini, S., Grethe, J.S.: Data Integration in the Biomedical Informatics Research Network (BIRN). In: Ludäscher, B., Raschid, L. (eds.) DILS 2005. LNCS (LNBI), vol. 3615, pp. 317–320. Springer, Heidelberg (2005)

6. Chen, I.A., Kosky, A.S., Markowitz, V.M., Szeto, E., Topaloglou, T.: Advanced Query Mechanisms for Biological Databases. In: 6th Int. Conf. on Intelligent Systems for Molecular Biology (1998)

7. Haas, L.M., Schwarz, P.M., Kodali, P., Kotlar, E., Rice, J., Swope, W.C.: DiscoveryLink: A System for Integrated Access to Life Sciences Data Sources. IBM Systems Journal 40(2), 489–511 (2001)

8. Do, H.H., Rahm, E.: Flexible Integration of Molecular-biological Annotation Data: The GenMapper Approach. In: Bertino, E., Christodoulakis, S., Plexousakis, D., Christophides, V., Koubarakis, M., Böhm, K., Ferrari, E. (eds.) EDBT 2004. LNCS, vol. 2992, pp. 811–822. Springer, Heidelberg (2004)

9. Shaker, R., Mork, P., Brockenbrough, J.S., Donel-son, L., Tarczy-Hornoch, P.: The Bio-Mediator System as a Tool for Integrating Biologic Databases on the Web. In: Workshop on Information Integration on the Web, IIWeb 2004 (2004)

10. Rahm, E., Bernstein, P.A.: A survey of approaches to automatic schema matching. The VLDB Journal 10(4), 334–350 (2001)

11. Leser, U., Naumann, F.: (Almost) Hands-Off Information Integration for the Life Sciences. In: Proceedings of the Conference on Innovative Data Systems Research, CIDR (2005)

12. Mullins, I., Siadaty, M., Lyman, J., Scully, K., Garrett, C., Miller, W., Muller, R., Robson, B., Apte, C., Weiss, S., Rigoustsos, I., Platt, D., Cohen, S., Knaus, W.: Data Mining and Clinical Data Repositories: Insights from a 667,000 Patient Data Set. Journal of Computers in Biology and Medicine (2005)

13. ISO 14721:2003, Blue Book. Issue 1. CCSDS 650.0-B-1: Reference Model for an Open Archival Information System, OAIS (2002)

14. Factor, M., Naor, D., Rabinovici-Cohen, S., Ramati, L., Reshef, P., Satran, J., Giaretta, D.L.: Preservation DataStores: Architecture for Preservation Aware Storage. In: 24th IEEE Conference on Mass Storage Systems and Technologies (MSST), San Diego, pp. 3–15 (2007)

Interpretation of History Pseudostates
in Orthogonal States of UML State Machines

Anna Derezińska and Romuald Pilitowski

Institute of Computer Science, Warsaw University of Technology
Nowowiejska 15/19, 00-665 Warsaw, Poland

Abstract. Inconsistencies and semantic variation points of the UML specification are a source of problems during code generation and execution of behavioral models. We discuss the interpretation of history concepts of UML 2.x state machines. Especially, history in complex states with orthogonal regions was considered. The clarification of this interpretation was proposed and explained by an example. The history issues and other variation points had to be resolved within the Framework for eXecutable UML (FXU). The FXU was the first framework supporting all elements of UML 2.x behavioral state machines in code generation and execution for C# code.

Keywords: UML state machines, statecharts, history, orthogonal regions, semantic variation points, UML code execution.

1 Introduction

The concept of state machine is used for the modeling of discrete behavior through the finite state transition system. A state machine is from its nature an automaton without memory. Its current behavior depends on a current active state (exactly - the configuration of active states) and the handling of occurring events. In UML state machines, this notion can be extended with the history mechanism [1]. A history pseudostate used in a composite state allows keeping trace of substates that were the most recently used between two entrances to the composite state. The history concept is very useful in creating concise, comprehensible but also very expressive models. For example, a suspension in business activities can be modeled with history [2].

The OMG standard leaves unspecified a number of details about the execution semantics of UML state machines. A great deal of work was devoted to the interpretation of state machines [2,3,4,5,6,7,8,9,10,11,12,13,14,15,16,17,18,19]. However, the history mechanism belongs to those advanced concepts that are often omitted or limited to simple cases. This situation can be observed in considerations about state machine semantics [18,19] and in tools dealing with state machines [20,21,22].

This paper addresses a problem of using history pseudostates in behavioral state machines. It is especially imprecise, when a composite state is an orthogonal state, and its region region belongs to the configuration of active states that

Y.A. Feldman, D. Kraft, and T. Kuflik (Eds.): NGITS 2009, LNCS 5831, pp. 26–37, 2009.

is currently associated with the history pseudostate. The problem, although is a theoretical one, has a practical motivation. Our goal was not to create another semantics, but clarify those inconsistencies or variation points that were indispensable for the interpretation of the UML models behavior. These interpretations were used in a code generation and execution framework of UML models FXU [23,24,25]. The framework is intended for building reliable applications using different modeling concepts, including orthogonal states and history.

In general, variation points can be useful and allow different interpretations according to requirements of a modeler. However, a clarification is necessary for some of them if we would like to interpret a state machine behavior. We discuss an open issue in the definition of UML state machines. An orthogonal state consists of several regions. A history pseudostate can be defined in one of the regions, and not directly in the state [1]. Reentrance of the orthogonal state via a history pseudostate can result in ambiguous behavior, because if one orthogonal region is active the rest of regions of the composite state should be also active. We propose the solution for this problem, treating this reentrance as if entering via "temporal fork". We suggest also other more general solutions, but they would require changes in the UML specification [1].

The rest of the paper is organized as follows. Section 2 summarizes briefly basic concepts of history in UML state machines. Further, different approaches to history concepts and UML state machine interpretation are summarized. Next, the proposed interpretation of entering history pseudostates in orthogonal states is explained and illustrated by an example. Section 5 describes application of the solution in the code generation and execution framework of UML classes and their state machines. Final remarks conclude the paper.

2   History Pseudostates in UML State Machines

This section recollects few basic concepts defined in the UML specification [1], which refers to history pseudostates.

State machines can be used to specify behavior of various model elements, e.g. classes. A state machine is represented by a graph of nodes interconnected by transition arcs. Transitions can be triggered by series of event occurrences. The behavior is modeled as a traversal of the graph. During this traversal, the state machine executes a series of activities associated with transitions or states.

A vertex, an abstraction of a node in a state machine graph, can be a state or a pseudostate. A state models a stable situation described by, usually implicit, invariant condition. There are following kinds of states: simple states, composite states and submachine states. A composite state either contains one *region* or is decomposed into two or more *orthogonal regions*. Each region has a set of mutually exclusive disjoint subvertices and a set of transitions. Any state enclosed within a region of a composite state is a substate of that composite state. It is called a *direct substate* when it is not contained by any other state; otherwise, it is referred to as an *indirect substate* which in turn own vertices and transitions.

If there are composite states, more than one state can be active at the same time. A set of all currently active states is called an *active state configuration*.

Once a simple state is active, all composite states that directly or indirectly encompass this state are also active. If any of these composite states is orthogonal, all of its regions are active. At most one direct substate is active in each region.

Event processing in state machines is based on the *run-to-completion* assumption. An event occurrence can be dispatched if the processing of the previous event occurrence is completed. Therefore, a state machine execution can be interpreted as a transition from one configuration of active states to another.

Pseudostates model different types of transient vertices in the state machine graph. Therefore, pseudostates are active only for the duration of transitions from one state configuration to another. The semantics of a pseudostate is determined by its kind attribute. Among ten different kinds of pseudostates we especially discuss here *deepHistory* and *shallowHistory*.

Both history pseudostates are used in composite states. They are used to "memorize recently active substates" during reentrance to a composite state. A composite state can have at most one deep history vertex. At most one transition may originate from the history connector to the *default deep history state*. The same conditions are true for shallow history vertex. Protocol state machines cannot have deep or shallow history pseudostates.

Shallow history entry has a following interpretation. Let us assume that a fired transition terminates on a shallow history pseudostate. If this is the first entry to the composite state enclosing this pseudostate or the most recently active substate is the final state, the *default shallow history state* is entered. In other situations, the active substate becomes the most recently active substate of the composite state. If the active substate determined by history is a composite state, then it proceeds with its default entry.

The rule of *deep history* entry is very similar to that of shallow history, except that it is applied recursively to all levels in the active state configuration below this one. In case the substates are only simple states (without further decomposition) both kinds of history are equivalent.

Basic notions of history are illustrated with an example (Fig. 1). Taking for the first time a transition outgoing from state *s1*, we come via the deep history pseudostate to its default history state, in this case state *s3*. State *s3* is enclosed in the composite state *s2*. Therefore, the first active configuration consists of states *s3* and *s2*. After occurrence of call event *Go*, a configuration of active states has three states (*s5*, *s4* and *s2*). Next configuration, i.e. after an occurrence of event *X*, has also three states (*s6*, *s4* and *s2*). Call event *Push* can be triggered when one of these three configurations is active. Because of the deep history pseudostate, a next taking of transition from *s1* results in one of the following situations: entering state *s3* (and *s2*), entering *s5* (and *s4*, *s2*) or entering *s6* (and *s4*, *s2*).

If there were a shallow history pseudostate in this example (instead of a deep one), only two configurations of states would be considered, i.e. (*s3* and *s2*) or (*s4* and *s2*). If state *s4* is entered as the first substate after history reentrance, its initial pseudostate (*s5*) will be visited.

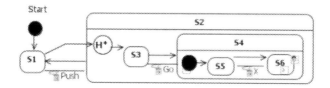

Fig. 1. Example 1 - a simple usage of a history pseudostate

3   A History Concept in Semantics of UML State Machines and UML Supporting Tools

The UML state machine is an object-based variant of Harel statecharts [4,5]. The UML specification [1] leaves certain semantic options open allowing semantic variation points. Other problems are inconsistency issues - aspects of state machines unintentionally not defined by the specification. Many different approaches were used in developing semantics of UML state machine [2,3,4,5,6,7,8,9,10,11,12,13,14,15,16,17,18,19] For the extensive lists (up to 2005) we can refer to [6,7]. However, in many approaches we could not find any interpretation to the history issues or the UML history concept is covered only partially (about 20% covers full or partially [6]).

The similar situation is in the tools dealing with state machines. Tools generating code from state machines consider often only a subset of elements of the UML 2.x specification (as surveyed in [25]). Many tools for model checking often ignore the history mechanism in statechart diagrams [2]. The most complete set of state machines' features is supported by the Rhapsody tool [22] of Telelogic (formerly I-Logix, currently in IBM) that generates code for C, C++, Ada and Java, as well as by FXU [23,24,25] (C# code). Tools building executable UML models [26,27] use different subsets of UML and there is no guarantee that two interchanged models will be executed in the same way. Therefore the OMG specification of semantics of a foundation subset for executable UML models (FUML) is prepared [28]. It defines a virtual machine for UML, but it considers only selected, mostly used UML elements, without history in state machines.

The detailed description of the Rhapsody semantics of statecharts was given in [5]. It explains system reaction on events and realization of compound transitions. The history mechanism is also discussed, but it is limited to standard description, similar to that given further in the UML specification and illustrated by a simple, typical example of a usage of a history pseudostate. The application of history in the context of orthogonal states is not considered.

Crane and Dingel [8] discussed several problems about modeling constructs and well-formedness constraints of statecharts. They compared three kinds of statecharts, classical Harel statecharts implemented in Statemate, state machines as described in the UML 2.0 specification, and another variant of object-oriented statecharts implemented in Rhapsody [22]. They showed that, although all solutions are very similar, there are differences influencing creation of a model and interpretation of its behavior. Although they did not discuss any problems of history pseudostates, it was pointed out that Rhapsody and Statemate support the deep and shallow history with considerable difference to UML 2.0.

In [9] semantics of UML statecharts is given in terms of interactive abstract state machines that exchange messages. In this approach semantic variation points, like choosing between conflicting transitions, are left open. Moreover, the statecharts are simplified, i.e. there are no deferred events, no history states and transitions may not cross boundaries within or out with composite states.

A structural operational semantics for UML statecharts was proposed in [10]. However, in this work, "last active" in explanation of a history pseudostate does not correspond to the exiting of the state.

In [11] semantics of UML 2.0 state machines was defined using the extended template semantics. Template semantics are defined in terms of an extended state machine that includes control states and state hierarchy, but not concurrency. Concurrency is introduced by composition operators. A UML orthogonal state is mapped to a composition of orthogonal regions using an interleaving operator. The history concept was omitted for the simplicity. Moreover, the approach of template semantics retains the semantic variation points that are documented in the OMG standard.

Another way of dealing with semantic variation points is proposed in [12]. The authors intend to build models that specify different variants and combine them with the statechart metamodel. Further, different policies should be implemented for these variants. Variants of time management, event selection and transition selection are shown. The history issues were not considered in this approach.

In [2] statecharts are converted into the formalism of Communicating Sequential Processes (CSP). Therefore, the tool supporting CSP can be used for model checking. In opposite to many similar approaches, the authors take into account the history mechanism which is useful for modeling suspension in business processes. Handling of history resulted in a more complicated semantic model than without history. Mapping statecharts into CSP was specified under several restrictions concerning history e.g. no shallow history was allowed, a history-bearing composite state must always be entered through its history. Using such restrictions they do not faced a problem considered in this paper.

In the contrast to many other papers, the work presented in [13] did not exclude history issues. However, the full current specification of statecharts in UML 2.x was not covered. Problems of entering of orthogonal states were considered only in such cases that can be simply resolved (see Sec. 4.1).

In [14] many ambiguities of UML behavioral state machines were discussed. The authors suggested that the concepts of history, priority, and entry/exit points have to be reconsidered. They noticed that history pseudostates belong to regions but on the other hand their semantics is defined for composite states containing them [1]. Further, the authors applied the semantics to a region containing a history pseudostate, whereas in our work the later case, i.e. the semantics of any composite state, especially orthogonal one, containing a history pseudostate was taken into account. The improvements of the history concept suggested in [14], and further developed by the authors in [15] did not cover the problem discussed here. The same is true for the semantics given in [16]. It resolves some problems identified in [14] assuming that allowed structure of state machines is more restricted than defined in [1]. For example, all regions have a unique initial pseudostate and unique default initial state.

Summing up, interpretation of history is either omitted, or considered for simple cases, or as in [14,15,16] partially explained under restricted usage.

4   Interpretation of Entering History Pseudostates

4.1   Problems with Orthogonal States

The problem of history interpretation refers to the state machine execution. It is similar to other unresolved problems of the UML state machines' specification [1], for example, priorities defining the order of firing of transitions, entering regions enclosed in an orthogonal state [23]. These problems should be coped while we would like to execute a UML model or at least have a unique interpretation of its behavior. The problem, we address here, is the order of entering states while entering a history pseudostate in a composite state with many regions.

In simple cases, as the example shown in Sec. 2, an interpretation of the state machine behavior is unique because only one substate is entered on the deepest level. Other states belonging to the current configuration of active states are the states encompassing this substate. For example, let us assume that the state machine was in state $s5$ when *Push* event occurred (Fig. 1). According to the deep history pseudostate, while executing transition outgoing from state $s1$, we enter state $s5$, its enclosing state $s4$ and its enclosing state $s2$. The entire semantics is exactly the same as if we had a direct transition outgoing from state $s1$ and targeting state $s5$. The order of actions is defined in the UML specification. It reflects the rule that nesting states are entered before the nested ones.

The problem arises when we deal with orthogonal states, i.e. having two or more regions. In general, it is possible to enter a whole orthogonal state, but also to target directly any substate belonging to one of regions of the state. According to the specification [1] if a composite state is active and orthogonal, all of its regions are active, with at most one substate in each region. Therefore other regions of the orthogonal state also should become active. The situation can be illustrated by an example (Fig. 2 reproduced from [13]). After following transition $t5$ (from state E to history pseudostate) the stored, active configuration of region $R1$ is taken into account. Region $R2$ should also become active, therefore we use default entry and enter state C.

But existence of the initial pseudostate in region $R2$ is not obligatory. Which state should be entered in region $R2$ if there is no initial pseudostate in it? In general the problem refers not only to the usage of history pseudostates. The UML specification [1] states that if a transition terminates on an enclosing state and the enclosed regions do not have an initial pseudostate, the interpretation of this situation is a semantic variation point. However, while dealing with executable state machines we should decide how resolve it.

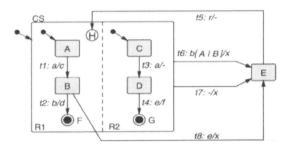

Fig. 2. A UML statechart diagram - example reproduced from [13]

One of solutions was shown in [23], but it referred to a case where no history pseudostate is present in the composite state. In that proposal the regions without direct entering transitions and without initial pseudostates were not entered and treated as completed. Such situations were not excluded from the model, but the behavior in some situations can be erroneous (for example cause a deadlock). However, it is not forbidden to create a state machine with an erroneous behavior. Such situation could be recognized with a warning, indicating that a model is probably ill-formed. It is also consistent with the specification [1].

The situation is different when a history pseudostate is used in the orthogonal state. The usage of history suggests that reentering of the composite state is intended by a developer. All the regions should be active after reentering of the composite state.

4.2 Interpretation

To resolve the above problem we propose the following interpretation.

Interpretation: Reentrance of a composite state via a deep history pseudostate can be treated equivalently as entering a temporal pseudostate fork. Transitions outgoing from this fork point to the deepest substates of the current active configuration associated with the history pseudostate.

Reentrance of a composite state in case of a shallow history pseudostate can be interpreted in the similar way as for a deep history. Therefore, reentrance a shallow history pseudostate in a composite state can be treated as if coming to a temporal pseudostate fork. A shallow history pseudostate "memorizes" only direct substates of its composite state. Transitions outgoing from the fork target these substates from the active configuration that refers to the shallow history.

Further interpretation of both kinds of history pseudostates is related to the behavior of orthogonal states. In general, we assume that if it is possible and it is not directly stated in a model, any model element should be not discriminated. Therefore, a concurrent behavior is assumed in the following cases:

1. orthogonal regions within a composite state,
2. enabled transitions included in a maximal set of non-conflicting transitions
3. different state machines, e.g. related to different classes.

The detailed description of these aspects is bmeyond the scope of this paper (see point 2 explained in [23]), but the resulting interpretation refers also for the considered history pseudostates. The transitions outgoing from the abovementioned temporal fork are enabled and non-conflicting. They can be fired concurrently. None of the substates of the active configuration will be discriminated.

It should be noted, that UML does not determine any model of concurrency. Therefore, a concurrent execution of transitions causes a risk of an uncontrolled access to common resources. A modeler should guarantee that any order of transitions is allowed or it should extend a model or its realization with some synchronization mechanism.

4.3   Example

The proposed interpretation and its consequences will be illustrated with an example. Let us consider a state machine shown in Fig. 3. The state machine has two simple states (s0, s10) and a composite state s1 owning three orthogonal regions. State s1 has four direct substates s2, s3, s6 and s7. Substates s3 and s7 are further decomposed into orthogonal regions.

The execution of the state machine starts with entering its initial pseudostate. Then the state s0 is entered, the only transition outgoing state s0 and targeting

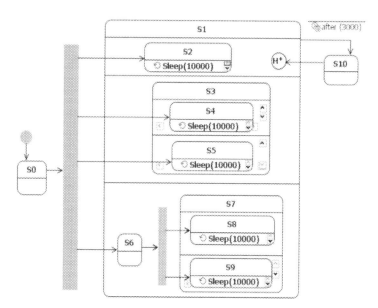

Fig. 3. A state machine with a history pseudostate used with orthogonal states

a fork pseudostate is fired. From the fork the execution comes to states *s1*, *s2*, *s3*, *s4*, *s5* and *s6*. Next, state *s6* is exited and three states *s7*, *s8* and *s9* are entered through a fork pseudostate. When 3000 ms passed after entering state *s1*, a configuration of active states consists of the following states *s1*, *s2*, *s3*, *s4*, *s5*, *s7*, *s8* and *s9*. All these states will be "memorize" in the context of a deep history pseudostate included in the composite state *s1*.

A time event triggers the transition outgoing from state *s1* to state *s10*. Afterwards, state *s10* is exited and a transition from *s10* to *s1* is fired. The transition enters the composite state *s1* through a deep history. Therefore the substates from the corresponding state configuration are entered. The execution comes to the deepest active substates i.e. *s2*, *s4*, *s5*, *s8*, *s9*. According to the given interpretation all the transitions are performed in parallel as if the substates were targeted from a fork pseudostate. Before entering those deepest substates, the appropriate composite states (*s1*, *s3* and *s7*) enclosing the substates are also entered. In result, the entire active state configuration is reentered.

4.4   Discussion

In the discussion of UML semantics given in [15,16], the region with no initial pseudostate is treated as ill-defined behavior. However, the same approach is considered in an alternative case, when any state of such region is not activated. This proposal will be similar to the solution shown in [23] for entering orthogonal regions, but without history. The solution shown in this paper requires a slightly modification of the given semantics. If an orthogonal state had a region with a history pseudostate, an active configuration should be collected not only for this region but also for other regions belonging to this state. They could be used while reentering this state if necessary. The rest of the semantics can remain unchanged.

In general, we suggest that the specification of history given in [1] should be extended, especially in case of orthogonal states. The solution could be even more radical than the above one shown in this paper. In the UML specification, a history is now defined in a region. However, if a state consists of many regions the situation is unclear even if we decide to use this or another interpretation.

It should be stressed that in the discussed example (Fig. 3) we cannot define a history pseudostate directly in orthogonal state *S1*. Although this could be seem a logical solution. The history must belong to a region, this means it can be defined in one of regions of *S1*, and it is not allowed to have few history pseudostates in different regions of *S1*. Summing up, we can propose also more general solutions than presented in previous subsections:

1. consider a history pseudostate in a state, therefore it should be valid for none or for all regions of an orthogonal state, or
2. allow and require usage of a history pseudostate in all regions of an orthogonal state, otherwise a model would be ill-formed, or
3. in case of orthogonal state associate always a history pseudostate with a fork pseudostate that transfers control to all regions of the state.

However, solutions given in points 1) and 2) require changing of the UML specification [1].

5   Application of the Interpretation of State Machines in the FXU Framework

Our interpretation of history pseudostates was applied in a code generation and execution tool called Framework for eXecutable UML (FXU) [24,25]. A distinguishing feature of FXU is the ability to handle every single element of behavioral state machines as defined by the current UML 2.x specification [1]. Therefore, for example, all possible ten kinds of pseudostates were taken into account.

The Framework for eXecutable UML (FXU) consists of two components:
- FXU Generator,
- FXU Runtime Library.

The FXU Generator transforms a UML 2.x model into the corresponding source code in C#. It analyses only classes and their state machines from the model. Similarly to other generators from UML to object-oriented languages, the FXU generator transforms the concepts from the class diagram into their counterparts in C#. The final executable application is build on the basis of three elements: this generated source code, an additional, directly written code (if necessary) and the FXU library.

The FXU Runtime Library hides logic of state machines and supports their execution. The library consists of classes that refer to the corresponding classes defined in the UML 2.x metamodel. Also for any kind of a pseudostate (e.g. a shallow history and a deep history) an appropriate class from the library implements its UML logic.

Many ideas encountering in state machines require parallel execution. Typical realization of such concepts is by multithreading. The FXU Runtime Library uses multithreading to deal with:
- processing of many state machines which are active at the same time,
- orthogonal regions working within states,
- processing of events.

Every state machine has its own queue which pools incoming events. Events can be both broadcasted and sent directly to the selected state machines. Events trigger transitions that have an active source state and when their guard conditions evaluate to true. If many transitions can be fired, transition priorities are used for their selection. Definition of transition priority provided by the UML specification is not sufficient to resolve all conflicts. Therefore, we had to extend the definition of transitions priority [23]. Selection of non-conflicting transitions is possible in any state configuration when the extended priority concept is applied. On the basic of our definition we dealt also with other problems related to entering and exiting of orthogonal states [23].

It should be noted that the problems of history pseudostates referred also to application of orthogonal states. Only having completed the interpretation of execution of orthogonal states, it was possible to provide a consistent behavior of UML state machines.

6 Conclusions

In the paper we discuss an interpretation of history pseudostates in the context of state machine behavior. In particular, transition to history pseudostates that refer to active configuration of substates included in orthogonal regions is not precisely determined in the UML specification [1]. The order of transitions to those substates can be viewed as a semantic variation point.

We propose to clarify this semantics by an interpretation as if the transition comes via a hypothetical fork pseudostate coming to all the appropriate substates. We assume also an interpretation of parallel execution of all orthogonal regions in all cases (regardless with or without history pseudostates). Supported by the interpretation of other variation points of transition firing as well as entering and exiting orthogonal regions [23], we obtain a consistent description of the behavior of a UML state machine. This is necessary to automate building an application that behaves equivalently to a given model.

The discussed approaches were implemented in the framework of code generation and execution FXU. The corresponding C# applications were created for several dozen of UML models. In the future work, we prepare other complex models implementing telecommunication problems. Capability of using advance state machine features, including history concepts, could be important in these applications.

References

1. Unified Modelling Language, http://www.uml.org
2. Yeung, W.L., Leung, K.R.P.H., Wang, J., Wei, D.: Modelling and Model Checking Suspendible Business Processses via Statechart Diagrams and CSP. Science of Computer Programming 65, 14–29 (2007)
3. van Langenhove, S.: Towards the Correctness of Software Behavior in UML. PhD Thesis, Ghent University, Belgium (2005)
4. Harel, D., Pnueli, A., Schmidt, J.P., Sherman, R.: On the Formal Semantics of State Machines. In: 2nd IEEE Symp. on Logic in Computer Science, pp. 54–64. IEEE Press, Los Alamitos (1987)
5. Harel, D., Kugler, H.: The Rhapsody Semantics of Statecharts (or On the Executable Core of the UML) (preliminary version). In: Ehrig, H., Damm, W., Desel, J., Große-Rhode, M., Reif, W., Schnieder, E., Westkämper, E. (eds.) INT 2004. LNCS, vol. 3147, pp. 325–354. Springer, Heidelberg (2004)
6. Crane, M., Dingel, J.: On the Semantics of UML State Machines: Categorization and Comparison. Technical Report 2005-501. School of Computing, Queens University of Kingston, Ontario, Canada (2005)
7. STL: UML 2 Semantics Project, References, Queen's University, http://www.cs.queensu.ca/home/stl/internal/uml2/refs.htm
8. Crane, M., Dingel, J.: UML vs. Classical vs. Rhapsody Statecharts: Not All Models are Created Equal. In: Briand, L.C., Williams, C. (eds.) MoDELS 2005. LNCS, vol. 3713, pp. 97–112. Springer, Heidelberg (2005)
9. Jurjens, J.: A UML Statecharts Semantics with Message-Passing. In: ACM Symp. on App. Comp., SAC 2002, pp. 1009–1013 (2002)

10. Beck, M.: A Structured Operational Semantics for UML Statecharts. Software and System Modeling 1(2), 130–141 (2002)
11. Taleghani, A., Atlee, J.M.: Semantic Variations Among UML StateMachines. In: Nierstrasz, O., Whittle, J., Harel, D., Reggio, G. (eds.) MoDELS 2006. LNCS, vol. 4199, pp. 245–259. Springer, Heidelberg (2006)
12. Chauvel, F., Jezequel, J.-M.: Code Generation from UML Models with Semantic Variation Points. In: Briand, L.C., Williams, C. (eds.) MoDELS 2005. LNCS, vol. 3713, pp. 54–68. Springer, Heidelberg (2005)
13. Jin, Y., Esser, R., Janneck, J.W.: A Method for Describing the Syntax and Semantics of UML Statecharts. Software and System Modeling 3(2), 150–163 (2004)
14. Fecher, H., Schönborn, J., Kyas, M., Roever, W.P.: 29 New Unclarities in the Semantics of UML 2.0 State Machines. In: Lau, K.-K., Banach, R. (eds.) ICFEM 2005. LNCS, vol. 3785, pp. 52–65. Springer, Heidelberg (2005)
15. Fecher, H., Schönborn, J.: UML 2.0 state machines: Complete formal semantics via core state machines. In: Brim, L., Haverkort, B.R., Leucker, M., van de Pol, J. (eds.) FMICS 2006 and PDMC 2006. LNCS, vol. 4346, pp. 244–260. Springer, Heidelberg (2007)
16. Lano, K., Clark, D.: Direct Semantics of Extended State Machines. Journal of Object Tecnology 6(9), 35–51 (2007)
17. Lam, V.S.W., Padget, J.: Analyzing Equivalences of UML Statechart Diagrams by Structural Congruence and Open Bisimulations. In: UML ZWI. LNCS, vol. 2185, pp. 406–421 (2001)
18. Lilius, J., Paltor, I.P.: Formalising UML State Machines for Model Checking. In: France, R.B., Rumpe, B. (eds.) UML 1999. LNCS, vol. 1723, pp. 430–444. Springer, Heidelberg (1999)
19. Hölscher, K., Ziemann, P., Gogolla, M.: On translating UML models into graph transformation systems. Journal of Visual Languages and Computing 17, 78–105 (2006)
20. Niaz, I.A., Tanaka, J.: Mapping UML Statecharts into Java code. In: IASTED Int. Conf. Software Engineering, pp. 111–116 (2004)
21. Knapp, A., Merz, S., Rauh, C.: Model Checking Timed UML State Machines and Collaborations. In: 7th Int. Symposium on Formal Techniques in Real-Time and Fault Tolerant Systems, pp. 395–414 (2002)
22. Rhapsody, http://www.telelogic.com/
23. Derezinska, A., Pilitowski, R.: Event Processing in Code Generation and Execution Framework of UML State Machines. In: Madeyski, L., et al. (eds.) Software Engineering in progress, Nakom, Poznań, pp. 80–92 (2007)
24. Pilitowski, R., Derezinska, A.: Code Generation and Execution Framework for UML 2.0 Classes and State Machines. In: Sobh, T. (ed.) Innovations and Advanced Techniques in Computer and Information Sciences and Engineering, pp. 421–427. Springer, Heidelberg (2007)
25. Pilitowski, R.: Generation of C# code from UML 2.0 class and state machine diagrams (in Polish). Master thesis, Inst. of Comp. Science, Warsaw Univ. of Technology, Poland (2006)
26. Mellor, S.J., Balcer, M.J.: Executable UML a Foundation for Model-Driven Architecture. Addison-Wesley, Reading (2002)
27. Carter, K.: iUMLite - xUML modeling tool, http://www.kc.com
28. Semantics of a foundation subset for executable UML models (FUML) (2008), http://www.omg.org/spec//FUML/

System Grokking – A Novel Approach for Software Understanding, Validation, and Evolution*

Maayan Goldstein and Dany Moshkovich

IBM Haifa Research Lab
Haifa University Campus, Mount Carmel,
Haifa, Israel 31905
{maayang,mdany}@il.ibm.com

Abstract. The complexity of software systems is continuously growing across a wide range of application domains. System architects are often faced with large complex systems and systems whose semantics may be difficult to understand, hidden, or even still evolving. Raising the level of abstraction of such systems can significantly improve their usability.

We introduce System Grokking[1] - a software architect assistance technology designed to support incremental and iterative user-driven understanding, validation, and evolution of complex software systems through higher levels of abstraction. The System Grokking technology enables semi-automatic discovery, manipulation, and visualization of groups of domain-specific software elements and the relationships between them to represent high-level structural and behavioral abstractions.

Keywords: software analysis, reverse engineering, modelling, patterns, software architecture.

1 Introduction

With the increase in software complexity, a need for improving the development process has emerged. Model Driven Architecture (MDA) [1] promotes the use of models as primary artifacts throughout the engineering lifecycle to leverage the work on higher levels of abstraction. Models are abstractions of entities or processes in a system. They are created to bring into focus different aspects of the entities or processes currently being analyzed. Models aim at hiding the unimportant data and emphasize what is noteworthy in connection with the current system analysis goal.

Models play a key role in the development process and have a tendency to become increasingly complex. In addition, models representing legacy systems may simply not

---

* This work is partially supported by the European Community under the Information Society Technologies (IST) program of the 6th FP for RTD - project MODELPLEX contract IST-034081. The authors are solely responsible for the content of this paper.
[1] From WhatIs.com: "To grok something is to understand something so well that it is fully absorbed into oneself". It was coined by science fiction writer Robert A. Heinlein in his novel "Stranger in a Strange Land".

Y.A. Feldman, D. Kraft, and T. Kuflik (Eds.): NGITS 2009, LNCS 5831, pp. 38–49, 2009.
© Springer-Verlag Berlin Heidelberg 2009

be available. Therefore, we need to be able to analyze systems and create models representing them. This provides a broader and more comprehensive picture of the system being examined. The process of system analysis is essential for the maintenance and qualitative development of systems throughout their lifecycle. It can also help detect possible violations of specific required characteristics.

The purpose of systems analysis may vary between different end users, depending on their viewpoint and intended use. In addition, end users prefer to study systems in an intuitive manner that fits their semantic perception. They define their logical abstractions over the software elements to categorize the information and then inspect the relationships between these abstractions to enrich their conceptual view. This process is usually incremental and iterative and requires tool support.

We therefore have introduced a novel approach for assisting system architects in improving the understanding, maintenance, quality, and evolution of large, complex systems. Based on this approach, we developed a system analysis technology known as the System Grokking technology.

This paper is organized as follows. The next two sections provide a detailed overview of our approach and the System Grokking technology. Section 4 exemplifies the technology's capabilities, while Section 5 reviews the state of the art. We provide conclusions and outline future directions of this research in Section 6.

2 Approach

Our approach is based on experience with various industrial partners and addresses three phases of the software development process: *understanding*, *validation*, and *evolution*. All these phases consist of a verity of complex tasks. Although some of the tasks can be automated, others may still require user input. Therefore, we believe that technology that supports these processes should allow intuitive visual interaction with users. We also suggest separating tasks into smaller incremental steps to deal with their often complex nature. To support this incremental and iterative user-driven process, we provide the user with visual modeling framework. The framework is based on a specially developed meta-model, and an extendable set of customizable analysis rules applicable through appropriate wizards.

Understanding an unfamiliar complex system is a difficult task, especially when the application was developed by different people over a long period of time. However, such an understanding is crucial to allow further development and maintenance of the application. During the understanding phase, the user is interested in discovery and visualization of the domain elements and the relationships between them. This is important as visual diagrams inspection is more intuitive than code inspection for many users. Although visualization is supported by different tools [15, 16], it often results in a rather basic transformation from low-level domain elements to models with only basic abstraction level.

To improve system comprehension, users need a set of mechanisms to detect and represent higher levels of abstraction. Grouping is one of the powerful mechanisms provided by the System Grokking technology to represent abstraction. Groups can be created manually or automatically through predefined set of rules and can describe any aspect of the system. The ability to classify software elements with user-defined or predefined types is another mechanism provided by our technology. It is especially useful for emphasizing semantic aspects of software elements.

Interactions between elements can be represented through typed relationships that can be manually specified by the user or discovered automatically by some of the analyses. The relationships can be grouped into derived relationships that allow the user to define high-level views of the system.

All these mechanisms provide the system architects with the ability to define valuable semantic abstractions to better understand and improve their systems. For instance, they may want to define semantic groups that represent application layers in the system. Users may further define semantic relations between these groups to represent layers dependencies. The understanding process may also include calculation of software metrics [5]. For instance, the users may be provided with information regarding the coupling level between the system's components.

The *validation* phase includes discovery and exposal of problematic relationships, such as cyclic dependencies between components, and detection of anti-patterns [7]. The user may want to specify a pattern (or use a predefined one) and check that the system does not validate this pattern. This is typically done in an iterative and incremental manner and involves interactions with the user. Upon detection of violation, the user may further want to pinpoint the offending components. Therefore, our approach supports logical operations such as union, intersection, and subtraction on groups of elements.

During the *evaluation* phase, users may want to simulate architectural improvements and changes at the model level. They may learn how the changes they seek to apply will impact the whole system or parts of it. Consequently, users may want to make additional changes, thus iteratively improving the system's architecture even more. Our approach assists users with their decision-making process by suggesting how to decompose the system into new components or how to arrange the components in different layers. Eventually, the changes can be applied to the underlying system by using refactoring [8] mechanisms or by code generation.

In the following section, we describe how our approach was implemented to support the aforementioned capabilities. We start by presenting a system overview after which we explain the semantic model used by our technology.

3 Technology Overview

The System Grokking technology serves as a visual manipulation framework for domain-specific software elements and the relationships between them. It was designed to provide end users with the variety of capabilities described in Section 2 and was partially implemented as a set of Eclipse plugins [2].

The System Grokking technology is illustrated in Figure 1. It shows an end user who wants to analyze a complex software system, addressed hereafter as the *target system*. Examples of target systems include UML [3] models, Java projects, etc. The target systems are represented by semantic models. Semantic models are created via domain-specific transformations following various semantic rules.

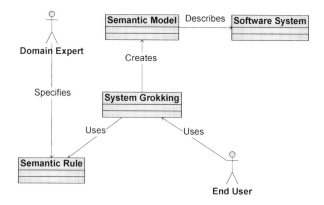

Fig. 1. System Grokking technology

Semantic models include several kinds of elements. These elements are generated by the System Grokking technology based on the semantic meta-model illustrated in Figure 2. The models are based on Eclipse Modeling Framework (EMF) [4] and are used for the visualization and persistency of the analysis results. There are four key types of elements in the meta-model: semantic rules, representors, groups, and relations.

Semantic rules, represented by instances of *SRule*, are configurable procedures that perform analysis and result in the creation of elements in the semantic model. They are specified by domain experts and are maintained for later exploitation by various end users. Semantic rules can be simple, such as the identification of all elements that depend on a given element. They can also offer more profound analyses. For example, they may support partitioning of the system into layers and verifying that the system complies with various design patterns.

Semantic representors, modeled by instances of *SRepresentor*, are local references to elements in the target systems. They may represent logical entities in the system such as classes, procedures, variables, etc. They may also represent physical resources such as files, folders, code, or text segments.

Semantic groups, represented by instances of *SGroup*, are sets of elements that share some common meaning. Examples of semantic groups include all the elements that belong to some application layer.

Semantic relations, represented by instances of *SRelation*, are used to express a relationship between elements in the semantic model. For example, semantic relations can be used to describe dependency relationships between two application layers, where each layer is represented as a semantic group.

Other types of elements shown in Figure 2 include the Semantic Element (*SElement*), which is the top meta-element in the semantic model class hierarchy. Elements that can be classified using semantic types (*SType*) are instances of *STypedElement*.

SDerivedRelation is a special kind of relationship, representing a relationship at a higher level of abstraction derived from a lower-level relationship. Consider for example two layers, *LayerA* and *LayerB*, which contain elements *A* and *B* correspondingly. If there is a relationship between *A* and *B*, then there is a derived relationship between *LayerA* and *LayerB*.

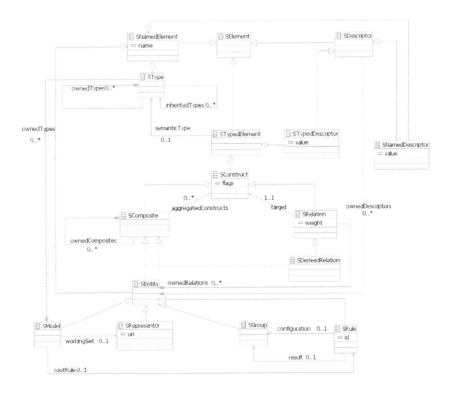

Fig. 2. Semantic meta-model

Finally, a Semantic Model (*SModel*) is a container for semantic abstractions of target systems. Each semantic model addresses a certain perspective of a system the user wants to inspect. It contains the relevant semantic representors that map the target system into the semantic model.

4   Usage Examples

In this section, we present a sample system implemented in C++ and exemplify some of the System Grokking technology's capabilities that are part of the understanding, validation, and evaluation processes.

4.1   Relationship Detection

The System Grokking technology supports the visual representation of software elements and the relationships between them through semantic models, as explained in Section 3. Consider, for example, the following C++ code snippet:

```
class SortAlgoInt {
public:
    virtual void sort(void ** arr, int len) = 0;
};

class FloatSortImpl: public SortAlgoInt {
public:
    void sort(void ** arr, int len) { //... }
    void sortDescending(void ** arr, int len){ //... }
};

class ComplexSortImpl: public SortAlgoInt {
public:
    void sort(void ** arr, int len) { //... }
};

class Factory {
public:
    enum DataType { FLOAT, COMPLEX };
    SortAlgoInt * getSortAlgorithm(DataType type) {
        if (type == FLOAT) return new FloatSortImpl();
        else return new ComplexSortImpl();
    }
};

int main() {
    void ** arr; int len;
    . . . //including initialization of arr and len
    SortAlgoInt * algo =
     (new Factory())->getSortAlgorithm(Factory::FLOAT);
    ((FloatSortImpl *) algo)->sortDescending(arr, len);
}
```

In the example, the two classes *FloatSortImpl* and *ComplexSortImpl* are used for sorting arrays of float or complex numbers, correspondingly. Both classes inherit from a common interface *SortAlgoInt* that contains a single method *sort*. Additionally, the class *FloatSortImpl* implements the *sortDescending* method, used to sort numbers in a descending order.

The code was developed based on a *Factory Method* [6] pattern. Therefore, the classes *FloatSortImpl* or *ComplexSortImpl* are instantiated by calling the factory method *getSortAlgorithm* of class *Factory*. The method receives as its parameter the type of data that needs to be sorted and returns the corresponding sorting algorithm implementation.

In the *main* method, an instance of *FloatSortImpl* is created by using the factory method. However, since the developer is interested in using the *sortDescending* method of this class, the casting mechanism is used to explicitly specify that the variable *algo* is of type *FloatSortImpl*.

The code snippet visualization is shown in Figure 3. There are five elements and four types of relationships between them. There are four classes (*Factory*, *SortAlgoInt*, *FloatSortImpl*, and *ComplexSortImpl*) represented by elements of type *CPPClass*. The *main* function is represented by an element of type *CPPFunction*. Since *FloatSortImpl* and *ComplexSortImpl* inherit from *SortAlgoInt*, there are two corresponding relations of type *CPPInheritance* between these classes. The numbers on the relations represent the weight of the relations. Since the implementation classes inherit only once from the interface class, the weight is set to 1.

The *main* function declares a variable of type *SortAlgoInt*. Therefore, there is a relationship of type *CPPDeclaration* between the corresponding elements in the diagram. Furthermore, since *main* function calls a method of *FloatSortImpl*, there is a relationship of type *CPPCall* between the corresponding elements.

Instantiations of objects of a certain type by calling their constructors are treated as *CPPCalls*. Therefore, the relationships between the elements representing the *Factory* class and the two implementation classes *FloatSortImpl* and *ComplexSortImpl* are of type *CPPCalls*. The instantiation of *Factory* from *main* method is treated identically. There is another method call from *main* to a method of *Factory*. Hence, the *CPPCall* relationship between the corresponding elements has the weight 2.

Since the *getSortAlgorithm* method's return type is a pointer to *SortAlgoInt*, there is a general relationship of type *Dependency* between *Factory* and *SortAlgoInt*.

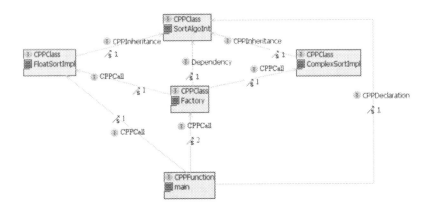

Fig. 3. Visualization of the system by the System Grokking technology

Once a model representing the system and the dependencies between the elements are created, system architects may use the results for further analysis. For instance, they may want to validate that the system has been implemented as intended. Since the system's implementation is based on Factory Method pattern, users may be interested in verifying that there are no violations of the pattern, as described below.

4.2   Design Patterns Representation and Detection

The *FactoryMethodPattern* can then be represented as a group with the semantic type *Pattern*, as shown in Figure 4. Additionally, we can create four groups that are contained by value [3] in *FactoryMethodPattern* and have the semantic type *Role: Interface, Implementation, Factory,* and *System*. The *System* group will represent all the elements that use the factory pattern elements and all the elements that are not related directly to the pattern.

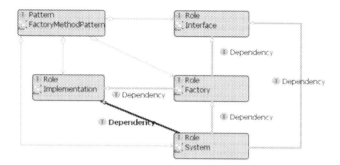

Fig. 4. Factory Method design pattern as implemented in the target system

The relations between the four groups are derived from direct dependencies (method invocations) between the elements comprising the groups. For instance, since normally *System* elements invoke methods defined in *Interface* element, there is a derived relationship between the two groups.

The System Grokking technology supports automatic grouping of a target system's elements into the pattern's elements according to their role in the pattern. For instance, it defines rules that use name-based queries to specify which system elements comprise the different pattern groups. In the example, the user may execute a semantic rule that will group all classes that have the *Impl* suffix in their class names in the *Implementation* group, and all classes that have the *Int* suffix in the *Interface* group, etc.

4.3 Investigation of Violations

Once the grouping step is completed, semantic rules can be used to discover violating elements. In our example, there is an offending relationship between *System* and *Implementation* that does not follow the Factory Method rules. This relationships is shown in Figure 4 as a thick black line with emphasized description (in bold).

The user may inspect the offending relationship by double-clicking the relation, as illustrated in Figure 5. The elements in the derived relations are shown as shortcuts to elements in the semantic model presented in Figure 3.

Fig. 5. The offending relationship's properties

The violation of the Factory Method pattern occurs due to the *sortDescending* method invocation made from within the *main* method. Users who want to correct this violation may apply the following changes to the code: add a new virtual method *sortDescending* to *SortAlgoInt* and implement the method in its child class, *ComplexSortImpl*. Consequently, users may remove the casting to the *FloatSortImpl* class. The following code snippet shows the modified parts of the original code:

```
class SortAlgoInt {
    //...
    virtual void sortDescending(void **, int) = 0;
};

class ComplexSortImpl: public SortAlgoInt {
    //...
    void sortDescending(void ** arr, int len) { //... }
};

int main() {
    //...
    SortAlgoInt * algo =
      (new Factory())->getSortAlgorithm(Factory::FLOAT);
    algo->sortDescending(arr, len);
}
```

In the example, the System Grokking technology helped improve the system's quality by detecting a defect and presenting it to the user. The user could further change the code to comply with the Factory Method pattern rules. The System Grokking technology can help end users modify various aspects of their systems. We refer to such modifications as the process of evolution.

4.4 Simulation of Architectural Changes

Assume now that the user wants to modify the corrected code by moving the classes *FloatSortImpl* and *ComplexSortImpl* into a new file. The user may simulate this by grouping the two classes in a *SortAlgorithmsImpl* group and by grouping the rest of the code in the *MainModule* group by subtracting the *SortAlgorithmsImpl* group from a group representing the original file, as shown in Figure 6.

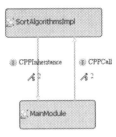

Fig. 6. Cyclic dependency between two groups of elements

The groups can be manipulated by using the various semantic rules. There are two derived relationships between the groups. One represents the inheritance relationships between *FloatSortImpl*, *ComplexSortImpl*, and *SortAlgoInt*. The other one represents instantiations of *FloatSortImpl* and *ComplexSortImpl* from within the *getSortAlgorithm* method of the *Factory* class. The user can invoke another rule defined in the

System Grokking technology to check the newly created groups for cyclic dependency. In our example, the cycle can be easily detected directly from the diagram.

To resolve the problematic dependency, users can use the related elements detection rule to create a new group of all elements in *MainModule* the *SortAlgorithmsImpl* group depends on (called *Interface*). They may further subtract the *Interface* group from the *MainModule* group, creating a new group *NewMainModule* that contains the remaining elements of the system.

Figure 7 shows the resulting model, without any cyclic relationship between the three groups. Consequently, the user may split the original code into three files and therefore improve the design of the analyzed system.

Fig. 7. Derived relationships between two groups of elements after correction

The next natural step would be to allow users to apply the changes they specified in a model to the target system's code. This is referred to in future work and is discussed in Section 6.

5 Related Work

There is a wide range of technologies that help coping with complex software systems. Most of them [10, 11] focus on representing different concepts of the systems that have some semantic meaning. Others [12-16] statically analyze or reverse engineer the code to create models that ease the understanding of the system.

Aspect-Oriented Software Development [11] is an approach to software development that uses a combination of language, environment, and method to assist programmers in the separation of cross-cutting concerns. Our technology bears some similarity to AOSD in that it allows the definition of some concerns using semantic groups and provides ways to inspect the relationships between them. However, the System Grokking technology concentrates more on semantic analysis of software rather than on separation or composition. Additionally, our technology is used to raise the level of abstraction and add expressiveness to models, while AOSD is used primarily to improve the software development process.

Domain ontology [10] is a formal organization of knowledge that represents a set of concepts within a domain and the relationships between those concepts. Domain ontology defines a formal abstraction over domain knowledge. The System Grokking technology can be configured by domain engineers to represent semantic abstractions over their domains. However, our technology provides additional capabilities. For example, it allows simulating architectural changes to the underlying system.

Numerous tools and analysis techniques exist and are being developed that address the issue of static analysis. These tools include many free open-source products as well as commercial offerings. Examples include IBM's RSAR [12] and Microsoft tools such as Visual Studio [13] and CodeAnalysis [14]. Some of these tools are designed to automatically support improvements in the software development process. For instance, RSAR has a rich set of analysis rules that can identify code-level issues. RSAR enables its users to exploit built-in analysis techniques and rules and create their own. Among the built-in analysis techniques are architectural pattern discovery (for JAVA), which includes design, structural, and system patterns. The System Grokking technology provides additional capabilities that allow, for example, simulating structural changes to the system.

Reverse engineering [15] is a process of discovering existing technology through an analysis of its structure, function, and operation. Most of the existing reverse engineering tools focus on reversing a static structure of the code, usually using modelling languages, rather than forming semantic abstractions. Converting the code to models without raising the level of abstraction merely transforms the description of the system from one formal language to another and thus does not necessarily reduce the complexity. The System Grokking technology can significantly leverage the reverse engineering process by providing a higher level of abstraction that addresses a user's perspective of the system.

Although there are several existing technologies in this area, to the best of our knowledge no single technology provides an end-to-end solution that supports the persistence, manipulation, visualization, and semi-automatic discovery of semantic abstractions over models as provided by the System Grokking technology. Moreover, none of the technologies supports the incremental and iterative user-driven approach suggested in this work.

6 Conclusion and Future Work

In this study, we introduced a novel approach that can assist in working with large and complex systems through the entire lifecycle. Specifically, we discussed the capabilities system architects might find very useful. We further presented the System Grokking technology, including its semantic meta-model, and showed how it can be used to support the end user with understanding, validating, and evaluating a target system.

The System Grokking technology allows users to investigate the existing semantic model in an incremental and iterative manner. It further allows users to discover and represent higher levels of abstractions and helps them follow architectural and semantic guidelines by identifying undesirable situations in their systems. In this work we exemplified how a predefined set of semi-automatic rules that can be used to, for instance, group system elements into patterns. We are currently working on implementing more sophisticated semantic rules that can, for instance, detect design patterns and anti-patterns automatically. One direction we are considering supporting is a visual, semantic-model-based pattern definition language and an appropriate rule-based mechanism for discovering these patterns. We are also developing algorithms for decomposing a system into logical layers, based on dependency analysis of components in the system.

End users may also want to define their specific analysis rules. Our technology has been designed to support this feature via a plug-in mechanism on the Eclipse platform [2]. We also plan to support OCL [9] queries for simpler rules.

The System Grokking technology allows the user to simulate changes in the semantic model that may be made to the system to improve and simplify its design. The next natural step would be to implement refactoring [8] mechanisms that would allow the user to apply the changes in the semantic model to the source code of the system. Some refactoring maybe easily implemented using the existing source code refactoring mechanisms, supported by development environments such as Eclipse [2]. Others may require new source code to be generated from the model. Techniques similar to those implemented in model-driven development environments, such as Rhapsody [16], can be used by the System Grokking technology for this purpose.

References

1. Model Driven Architecture, http://www.omg.org/mda
2. Eclipse Open Source Community, http://www.eclipse.org
3. Larman, C.: Applying UML and Patterns: An Introduction to Object-Oriented Analysis and Design and Iterative Development, 3rd edn. Prentice Hall, Englewood Cliffs (2004)
4. Eclipse Modelling Framework, http://www.eclipse.org/modeling/emf/
5. Kan, S.H.: Metrics and Models in Software Quality Engineering, 2nd edn. Addison-Wesley, Reading (2002)
6. Gamma, E., Helm, R., Johnson, R., Vlissides, J.M.: Design Patterns: Elements of Reusable Object-Oriented Software. Addison-Wesley, Reading (1994)
7. Brown, W.J., Malveau, R.C., Mowbray, T.J.: AntiPatterns: Refactoring Software, Architectures, and Projects in Crisis. Wiley, Chichester (1998)
8. Fowler, M., Beck, K., Brant, J., Opdyke, W., Roberts, D.: Refactoring: Improving the Design of Existing Code. Addison-Wesley, Reading (1999)
9. OMG, UML 2.0 OCL Specification, adopted specification, ptc/03-10-14
10. Djurić, D., Gašević, D., Devedžić, V.: Ontology Modeling and MDA. Journal of Object Technology 4(1) (2005)
11. Chuen, L.: Aspect-Oriented Programming in Software Engineering, http://users.wfu.edu/chen14/aop.pdf
12. IBM Rational Software Analyzer, http://www.ibm.com/developerworks/rational/products/rsar/
13. Microsoft Visual Studio, http://www.microsoft.com/visualstudio/
14. Microsoft Code Analysis, http://code.msdn.microsoft.com/codeanalysis
15. Hassan, A., Holt, R.: The Small World of Software Reverse Engineering. In: 11th Working Conference on Reverse Engineering, pp. 278–283. IEEE Computer Society, Washington (2004)
16. Telelogic Rhapsody, http://www.telelogic.com/products/rhapsody/index.cfm

Refactoring of Statecharts

Moria Abadi[1] and Yishai A. Feldman[2]

[1] Tel Aviv University
[2] IBM Haifa Research Lab

Abstract. Statecharts are an important tool for specifying the behavior of reactive systems, and development tools can automatically generate object-oriented code from them. As the system is refactored, it is necessary to modify the associated statecharts as well, performing operations such as grouping or ungrouping states, extracting part of a statechart into a separate class, and merging states and transitions. Refactoring tools embedded in object-oriented development environments are making it much easier for developers to modify their programs. However, tool support for refactoring statecharts does not yet exist. As a result, developers avoid making certain changes that are too difficult to perform manually, even though design quality deteriorates.

Methodologically, statecharts were meant to enable a systems engineer to describe a complete system, which would then be refined into a concrete implementation (object-oriented or other). This process is not supported by object-oriented development environments, which force each statechart to be specified as part of a class. Automated tool support for refactoring statecharts will also make this kind of refinement possible.

This paper describes a case study that shows the usefulness of refactoring support for statecharts, and presents an initial catalog of relevant refactorings. We show that a top-down refinement process helps identify the tasks and classes in a natural way.

1 Introduction

Statecharts [1] generalize finite-state machines by adding hierarchy, concurrency, and communication, thus enabling concise specifications of reactive systems. An example of concurrency is the state `Operating` in Figure 1, which has three concurrent substates, depicted one above the other and separated by horizontal lines. These describe components that may execute concurrently. The top component consists of a single state, called `Transmitting`. This is a compound state, whose behavior is hidden in this figure. The details appear in Figure 7; these could also have been shown in Figure 1, depending on the desired level of detail. Figure 7 shows two states, `DataTransmitting` and `VoiceTransmitting`, which are detailed in the same figure. Each of these has two parallel components. Every event in a statechart is broadcast and can be sensed in any currently-active state.

The statchart formalism can be used by system engineers to specify the behavior of complete reactive systems or parts of such systems, regardless of implementation details. For example, the behavior of an avionics system could be

Y.A. Feldman, D. Kraft, and T. Kuflik (Eds.): NGITS 2009, LNCS 5831, pp. 50–62, 2009.
© Springer-Verlag Berlin Heidelberg 2009

specified using one or more statecharts. The system would then be implemented using a combination of hardware and software components, where component boundaries do not necessarily correspond to states in the specification. The software can be implemented in a mixture of technologies, both object-oriented and others. The first tool for developing statechart specifications, STATEMATE [2], was not tied to a particular implementation paradigm.

More recent tools, such as Rhapsody [3,4], require statecharts to describe the implementation of a single class. This implies that statecharts can no longer be used to specify a complete system, only those parts of it that have already been defined as separate classes. In particular, much of the power of the statechart formalism to describe concurrency is not used, since there is not much parallelism in the behavior of a single object, and parallel states are mostly reduced to describing non-determinism (or implementation freedom) rather than true parallel behavior.

Furthermore, the class-based statechart specification makes it difficult to refactor the object-oriented structure of the system. A refactoring that changes the responsibilities of classes would often require a corresponding refactoring of the statecharts that specifies class behaviors. In the absence of tools for refactoring statecharts, developers avoid those object-oriented refactorings that would require statechart changes, negatively impacting the overall system design.

We claim it is necessary to create a formal catalog of statechart refactorings in order to build statechart-refactoring tools, in the spirit of existing code-refactoring tools. This is necessary in order to support both the systematic refinement of implementation-independent statechart specifications of systems into hardware and software components, and the ongoing refactoring of existing systems to improve their designs in the presence of changing requirements.

We have performed an initial case study of the refinement of a specification of a communications system, based on commercial designs. Our case study has shown that the systematic refinement process of a system-wide statechart specification naturally leads to the identification of both classes and tasks. This method seems to be much more beneficial for the implementation of reactive systems than identifying classes based on use cases.

Automation of these refactorings requires careful attention to their validity, so as not to change the specified behavior. This involves the source and target states, conditions, and entry and exit actions. Particularly important are data dependencies between states and statecharts, which are usually not obvious in the statechart specification. For example, when extracting a statechart it is necessary to identify data flow between the two statecharts and to pass the required parameters and results between them.

2   Refactoring Statecharts: An Example

This section shows several excerpts from our case study, demonstrating the use of some key refactorings on statecharts and how classes and tasks are identified as part of the refactoring process. This scenario shows how top-down refinement of

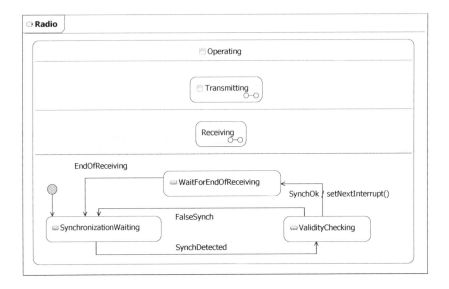

Fig. 1. Main `Radio` statechart

a system can be carried out using a set of statechart refactorings. However, it is also possible to use the same refactorings to modify existing systems to change their design and move functionality, including parts of statecharts describing reactive behavior, between classes.

The example starts with a high-level design of a communications system, expressed as a system-wide statechart (Figure 1). This statechart belongs to the class `Radio`, but describes much more than the behavior of this single class. At this level, the system is described as having three parallel components. The first two, `Transmitting` and `Receiving`, are considered primitives at this level. The third scans every frequency in a given range for a synchronization word. When the synchronization word is received at some frequency, the receiving process will start listening at that frequency (by sensing the *SynchOK* event).

The last component seems a good candidate for extraction into a separate task (and class), since it is executed in parallel. This is the Extract Statechart refactoring, which removes part of a statechart into its own class. The new class, `Synchronizer`, is shown in Figure 2. Once the synchronizer has been extracted to its own class, it is completely removed from the `Radio` class, and the new class is made into a separate task.

Of course, the previous behavior needs to be preserved. This implies preservation of events as well as data flow across classes. Events now need to be sent as messages, since different statecharts cannot sense each other's events directly. In this example, note the changed label of the transition from `ValidityChecking` to `WaitForEndOfReceiving`. The event *SynchOK* is now generated by the statechart of class `Synchronizer`, which has to send it to the *radio* object of the

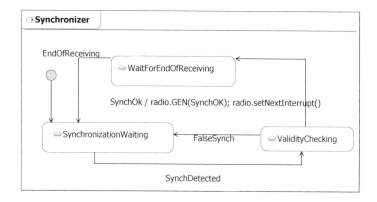

Fig. 2. Synchronizer statechart

class from which the statechart was extracted, so that the `Radio` statechart can react to it. (We use the Rhapsody convention [3] that events are sent by the `GEN` method.) In addition, the order of events may need to be preserved for correctness; we return to this issue in Section 4.

We now continue the refinement of the `Receiving` state. The first step is shown in Figure 3. It consists of two orthogonal components. The one labeled A (top of Figure 3) waits for the event *SynchOK* from the synchronizer, and then stabilizes the hardware to the appropriate frequency. (Note that the data flow of the frequency from the synchronizer to the radio is not shown in any statechart, but the design tool must keep track of it for proper code generation.)

When the second component (labeled B, bottom of Figure 3), which assembles packets, issues the *PacketReady* event, the first component decodes the packet header to determine whether to process the packet as voice or data.

The two components of `Receiving` deal with different data, at different levels of abstraction, and both need to be executed in parallel. This again calls for an Extract Statechart refactoring to create a separate class (and task) for the second component; the class is called `PacketReceiver` (see Figure 9). In addition, the first component deals with two separate issues: decoding the header, and processing the packet contents. In order to separate the former, it is first necessary to group all relevant states into a single super-state. This is the Group States refactoring [5], whose results are shown in Figure 4.

Now it is possible to extract this state into a separate class, called `Header`. Note that this class is not a separate task, since it does not execute in parallel to `Receiving`. As mentioned above, a refactoring tool must keep track of how these refactorings affect other parts of the system. For example, once header processing is extracted into its own class, `PacketReceiver` must be modified to send the *PacketReady* event to the header instead of the radio.

This case of Extract Statechart is more complicated than the previous one. When extracting a complete parallel component, nothing needs to be left behind in its place. In this case, however, we want to extract a non-parallel state. In

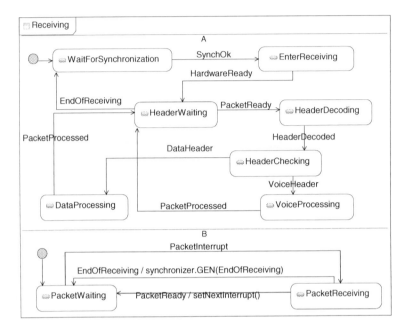

Fig. 3. Receiving statechart

order to do this, it is necessary to leave behind a primitive state, and connect all incoming and going transitions properly. In general, it is necessary to add an initial state to the extracted statechart; in this example, it is called Idle (see Figure 5). This state waits for the *Activate* event, which is now sent to it explicitly by actions on all transitions that entered HeaderWaiting in the grouped state of Figure 4. Similarly, the transitions that exited the grouped state HeaderProcessing in Figure 4 now go back to the new Idle state, sending their former actions to radio. The result is shown in Figure 6.

In this particular case, however, the additional Idle state is not really necessary, since the state HeaderProcessing being extracted already has an initial state HeaderWaiting that can serve this purpose. The developer can merge both states into one using the Merge States refactoring [6]. This new state will have a self-transition labeled with *Activate*, which is now redundant. The Remove Self-Transition refactoring can then be used to remove this transition, as well as all places where the event *Activate* is generated.

We now turn to the refinement of transmission processing. The first step is shown in Figure 7. This statechart waits for a request to transmit (identified by the event *Request To Transmit*) to start transmission, which can happen in one of two ways according to the type: voice or data. VoiceTransmitting and DataTransmitting have a very similar structure, although the details (such as how the data is encoded and laid out in the packet) are different. In both cases there is a component (shown on top) that prepares packets for transmission and

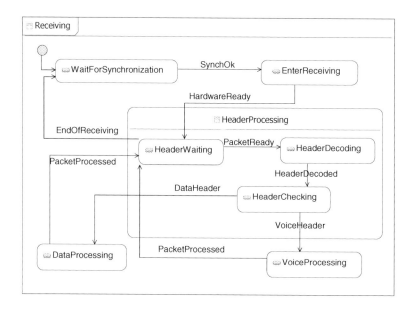

Fig. 4. Grouping header processing states

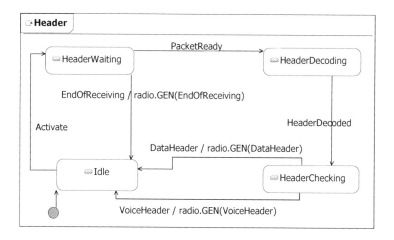

Fig. 5. Extracted header processing

transmits them, and a parallel component (at the bottom) that manages the buffers (two for each type of transmission). (The events *AReady* and *BReady* are common to the two bottom components; this is legal because they are not parallel. Because of the similarity between the states, this is a common idiom.)

Since buffer management needs to happen in parallel to the other transmission tasks, the buffer-management components should be extracted into separate

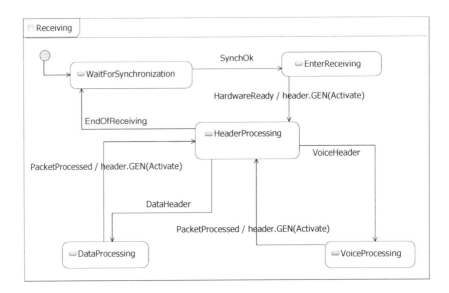

Fig. 6. Statechart for `Receiving` after extraction of header

tasks (and classes), using another Extract Statechart refactoring. Since they
have identical structure, it is natural to group them under a common superclass,
called `Recording` (see Figure 9). We call this the Pull-Up Statechart refactoring.
We can use the same refactoring on the remainder (top part) of the two transmitting states, resulting in the `PacketTransmitter` common superclass, whose
statechart is shown in Figure 8.

The class diagram in Figure 9 shows the final decomposition resulting from
this scenario. As described above, decisions about task and class decomposition
arise naturally out of the refactoring process.

3   Statechart Refactoring Catalog

This section describes the initial catalog of statechart refactorings used in our
case study (not all of which were shown above). Some of these have been described before in other contexts. Like object-oriented refactorings [7], it is useful
to have refactorings in pairs, each element of which can be used to undo the effects of the other. Below we describe at least one refactoring of each pair, usually
the one used in our scenario. Each refactoring has preconditions on its applica-
bility, and may need to be aware of and transform statecharts in other classes
as well as features like data flow that are not shown explicitly in the statechart.

The most useful refactoring is Extract Statechart, which is the core of the
top-down refinement methodology. It extracts a state (typically a composite one)
from a statechart and moves it to another class. One example that appeared in

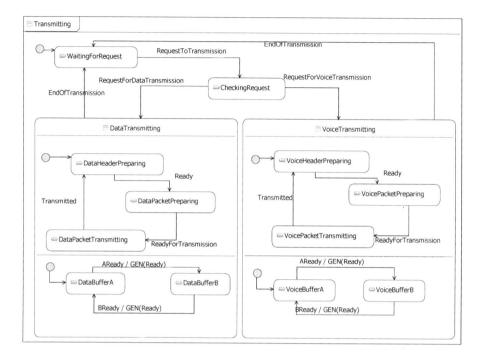

Fig. 7. Transmitting statechart

the scenario above was the extraction of the synchronizer from the `Radio` class; another was the extraction of header processing from `Receiving`.

Extracting a state makes some events non-local; these need to be replaced by explicitly sending the events to the appropriate object. Also, if there are external messages that the extracted state listens to, they must now be routed to the new object, in addition to or instead of the current one. This occurred in our scenario when we extracted the header-processing portion of the `Receiving` state into its own `Header` class, necessitating the routing of the *PacketReady* event to this new class instead of to the `Radio` class.

Data flow is crucial to the correct behavior of the system, but is not shown explicitly in the statecharts. A statechart refactoring tool must keep track of data flow that crosses object boundaries, so as to enable the correct code to be generated to route it. Standard data-flow algorithms can be used for sequential states, but they need to be modified to support data flow between parallel states.

In addition, a refactoring tool must also check syntactic preconditions on the proposed refactoring. For example, it is impossible to extract a state if multiple transitions enter different substates.

The above considerations on automation of Extract Statechart also apply to other refactorings as well (to varying degrees).

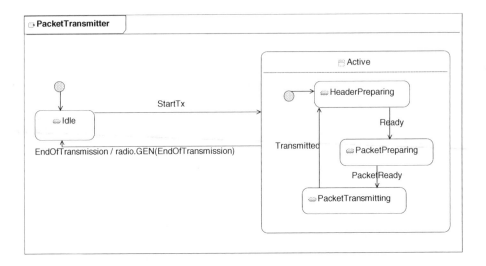

Fig. 8. Packet transmitter, common superclass for voice and data transmitters

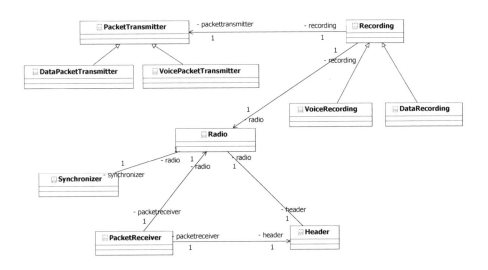

Fig. 9. Implementation classes in the resulting design

In order to extract several states simultaneously, they first need to be grouped into a single state, as we did in preparation for extracting the header processing (Figure 4). This is achieved by the Group States refactoring [5]. The reverse refactoring is called Flatten States [5].

The Merge States [6] refactoring can be used to combine two states that have identical behaviors into a single state. This, too, may be the result of previous modifications. One example of this case appeared in the extraction of header processing above. The reverse refactoring can be used to split one state into two identical parts, which can then evolve separately.

As happened in the example, when merging states it is possible for redundant self-transitions to emerge. If these have no associated actions, they can be removed. If, in addition, the activating event for these self-transitions is not used elsewhere, it can be removed as well. All these checks and actions are performed by the Remove Self-Transition refactoring.

It is also possible to split and merge transitions. Suppose a certain composite state has an exiting transition labeled with the event A that enters a substate HandleA; this implies that whenvever A occurs in any substate, state HandleA is entered (and any associated actions performed). However, during development it may become necessary to change the target or actions for some of the substates. In that case, the Unfold Outgoing Transitions [5] refactoring can be used to replace the single transition with multiple identical transitions, one for each substate. These can then be individually modified as necessary.

In preparation for Extract Statechart, it may be necessary to add states into a composite state, or remove substates from it. These transformations are achieved by the Insert State into Composite State and Extract State from Composite State refactorings [5].

Sometimes developers put a series of actions on a single transition; these can be combined into a single method by the Extract Transition Method refactoring. Similarly, a conditional in the action of a transition can be replaced by two separate transitions with different conditions by the Extract Condition refactoring. These refactorings can significantly improve the understandability of the statechart.

Finally, we saw the Pull-Up Statechart refactoring used in the scenario above to create an inheritance hierarchy when two statecharts with identical structure are found, moving common behavior to the superclass.

4   Refactoring Challanges

There are several types of constraints that a refactoring tool for statecharts needs to maintain. First, it needs to check syntactic preconditions on various transformations. For example, when inserting a state B into an existing state A, the tool must check that any transition exiting A is compatible with the transitions exiting B, and that any entry and exit action of A is compatible with the corresponding action of B, if any.

As shown in the scenario, the tool must also keep track of data flow between classes and statecharts. An Extract Statechart refactoring may have an effect on other statecharts, which are neigher in the original class nor in the new one. For example, if the receiving component is extracted from the radio into its own class, messages carrying events such as *SynchOK* now need to be routed to the new receiving class instead of, or in addition to, the radio.

The most difficult challenge for a statechart-refactoring tool is preserving the original behavior in spite of the different event-handling semantics in a single

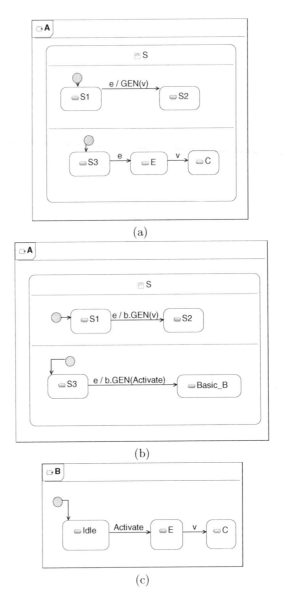

Fig. 10. Change of semantics from single statechart to multiple ones: (a) original statechart; (b) after removal of E and C; (c) extracted statechart

statechart and between classes in an object-oriented framework. Consider, for example, the statechart in Figure 10(a). If we group the state E and C and then extract the resulting state, the original statechart will become the one in Figure 10(b), while the extracted part will appears in Figure 10(c).

In the original statechart, if the event e occurs when the statechart is in its initial configuration {S1, S3}, the top component will move from S1 to S2, and simultaneously the bottom component will move from S3 to E. The former transition will generate the event v, which will then cause the bottom component to move from E to C. The final configuration will therefore be {S2, C}.

In the new statechart, the event e will cause the transitions from S1 to S2 and from S3 to Basic_B. Two new events will be sent to the object b: v and *Activate*. Because the semantics of object-oriented message passing is asynchronous, the order in which these events will arrive is unspecified. If *Activate* precedes v, the statechart of the newly-extracted component will move to E and then to C, ending in the same states as the original statechart. However, if they appear in the opposite order, the first will be ignored, and the final configuration will be {S2, E}. This behavior was not allowed by the original specification, and should not be allowed by any behavior-preserving transformation. Preventing these undersired behaviors is an important topic for future research.

5   Conclusions and Future Work

The development scenario shown in Section 2 demonstrated the need for automated refactoring tools to support both top-down refinement of a system-wide statechart into an object-oriented implementation, as well as for making modifications to existing designs. Section 3 presented an initial catalog of statechart-related refactorings, and briefly discussed what would be required for their automation. The information that the tool needs to keep track of includes syntactic preconditions, data flow between different classes, and events exchanged through messages between different classes.

The most novel refactorings in our catalog, Extract Statechart and Pull-Up Statechart, make non-local changes and require system-wide analysis. The automation of these refactorings would be particularly useful, as they are the most difficult to do manually.

Our ultimate goal is the automation of statechart-related refactorings to support the refinement and evolution of reactive systems from system design to implementation. These capabilities will complement the code refactoring tools available today. In order to do this, we intend to add more case studies like the one presented in this paper, in order to identify the required refactorings. These will be formalized, together with their preconditions and detailed effects. This catalog will then form the basis for the implementation.

References

1. Harel, D.: Statecharts: A visual formalism for complex systems. Science of Computer Programming 8, 231–274 (1987)
2. Harel, D., Politi, M.: Modeling Reactive Systems with Statecharts: The Statemate Approach. McGraw-Hill, Inc., New York (1998)
3. Harel, D., Gery, E.: Executable object modeling with statecharts. IEEE Computer, 31–42 (July 1997)

4. Harel, D., Kugler, H.: The Rhapsody semantics of statecharts (or, on the executable core of the UML). In: Ehrig, H., Damm, W., Desel, J., Große-Rhode, M., Reif, W., Schnieder, E., Westkämper, E. (eds.) INT 2004. LNCS, vol. 3147, pp. 325–354. Springer, Heidelberg (2004)
5. Sunyé, G., Pollet, D., Le Traon, Y., Jézéquel, J.M.: Refactoring UML models. In: Gogolla, M., Kobryn, C. (eds.) UML 2001. LNCS, vol. 2185, pp. 134–148. Springer, Heidelberg (2001)
6. Mens, T.: On the use of graph transformations for model refactoring. In: Lämmel, R., Saraiva, J., Visser, J. (eds.) GTTSE 2005. LNCS, vol. 4143, pp. 219–257. Springer, Heidelberg (2006)
7. Fowler, M.: Refactoring: Improving the Design of Existing Code. Addison-Wesley, Reading (2000)

Towards Health 2.0: Mashups to the Rescue

Ohad Greenshpan[1,2], Ksenya Kveler[1], Boaz Carmeli[1],
Haim Nelken[1], and Pnina Vortman[1]

[1] IBM Haifa Research Lab
[2] Tel-Aviv University

Abstract. Over the past few years, we have witnessed a rise in the use of the web for health purposes. Patients have begun to manage their own health data online, use health-related services, search for information, and share it with others. The cooperation of healthcare constituents towards making collaboration platforms available is known today as Health 2.0. The significance of Health 2.0 lies in the transformation of the patient from a healthcare consumer to an active participant in a new environment. We analyze the trend and propose mashups as a leading technology for the integration of relevant data, services, and applications. We present *Medic-kIT*, a mashup-based patient-centric Extended Personal Health Record system, which adheres to web 2.0 standards. We conclude by highlighting unique aspects that will have to be addressed to enable the development of such systems in the future.

1 Health 2.0

During the past few years, a trend has emerged towards using the web for health purposes. Millions of users who once connected online mainly via email discussion groups and chat rooms [1] now upload information, mine health-oriented web sites, and build more sophisticated virtual communities to share generated knowledge about treatment and coping. At the same time, traditional web sites that once offered cumbersome pages of static data are now providing data and tools to help users peruse the most relevant and timely information on health topics. This trend, in which patients, physicians, and providers focus on healthcare value while using current trends in IT and collaboration is known today as *Health 2.0* [2] In a few years from now, large volumes of health data will be available electronically. By then, various constituents in the health ecosystem will demand on-line access to that data from anywhere, exchanging information and seeking new opportunities to improve health. This situation will cause a revolution in the way healthcare players interact and will result in a sharp increase in awareness, knowledge, and the quality of treatment. Similar to the current natural evolution of other domains in *web 2.0*, the focus will shift from the healthcare providers and payers (i.e., insurance companies) to the patients and their communities. Consequently, new business models that do not exist today will arise, and, with them, requirements for a better means to conduct data management, exploration, and sharing. Even though Health 2.0 has much in common with web 2.0, Health 2.0 will have its own unique characteristics, as discussed in this paper.

From an IT perspective, collaboration in health began with the creation of Electronic Medical Record systems (EMRs). These systems are repositories of an individual's

Y.A. Feldman, D. Kraft, and T. Kuflik (Eds.): NGITS 2009, LNCS 5831, pp. 63–72, 2009.

health information, but are managed by their caregivers. Such systems are also used to pass information between caregivers. Pratt et al. [3] claim that despite the clear value of EMRs, they have a strong tendency to fail during phases of development and deployment due to being large and complex systems. They proposed that reconsidering EMRs from a Computer Support Collaborative Work (CSCW) perspective could provide valuable insights, leading to more successful EMRs and better medical information systems in general. The next natural development of EMRs was the Personal Health Record system (PHR). Like EMRs, PHRs enable information management, but are designed to be managed by the patients themselves and not by a central authority. The goal of PHRs is not just to store data, but to combine data and knowledge with software tools, which helps patients become active participants in their own care. The spread of PHRs is growing but still limited, despite the variety of potential benefits they offer both patients and caregivers. Tang et al. [4] claim that the widespread adoption of PHRs will not occur unless they provide a perceptible value to users, and are easy to learn and use.

The scope of the term *health* has expanded over the years, and now contains aspects such as environmental factors and lifestyle that were not in focus for many years. In addition to being more aware of their health than ever before, patients expect to control their care and treatment and want to share information with others. Thus, PHRs will have to be adjusted to contain new sets of components, satisfying these evolving needs. We call this new generation of PHRs *Extended PHRs (xPHRs)*.

This paper discusses three important collaboration aspects that must be addressed when designing such xPHRs: i) System flexibility that fits the needs of the "Long Tail", ii) Enabling patient-centric collaboration with the caregiver, and iii) Supporting easy interaction between patients and their communities. Other important aspects, such as security, privacy, and data persistence, are beyond the scope of this paper.

1.1   Flexibility for the Long Tail

Patients in the evolving Web 2.0 era, named as Health 2.0 patients in this paper, each with his/her individual needs, are likely to demand personalized systems enabling the management of increasing volumes of data and supporting heterogenous services. We expect this trend of "the long tail" distribution of requirements [5] to be similar to other domains on the web. For such systems, as we call xPHRs, to meet the requirements of each individual, they will need to be flexible, containing the type of data the user needs, and allowing the user convenient access to the data. The xPHRs will also have to be dynamic, customizable, and highly available at any time and place. To cope with complex data and flows, they will need to enable high interoperability between services supplied by various providers, with adherence to common standards, both healthcare (e.g., HL7, IHE) and data (e.g. RSS, ATOM, Web Services).

1.2   Cooperative Patient-Centric Care

In the traditional healthcare process, patients are mostly passive. They usually initiate the treatment process after identifying some abnormal condition and then let their physicians lead them towards successful treatment. With the development of the Internet, new means of communication between patients and their caregivers have begun to appear, as described thoroughly by [6]. The evolvement of PHRs allows patients to monitor the treatment process more closely and to collect relevant information about it. With the development of xPHRs, patients will desire further control of the treatment process. They are likely to expect easy interaction with multiple physicians by sharing their information, consulting, interacting, and being involved in decisions related to further treatments [7].

1.3   Patients and Their Communities

Since the first Usenet news sharing programs were created around 1979, online communities have co-evolved with the growth in computer networking [8]. Today, 30 years later, people share a great deal of information (e.g., news, music, pictures) in online communities, benefit from the presence and activities of others, and contribute knowledge and support. Similar to the clear trend in other domains, users belonging to disease-centric communities use such communities to share information, experiences, and provide psychological support. Currently, such communities are based on free and unorganized platforms, such as wikis or forums, and are mainly used to exchange information and knowledge, as described by [9]. McGee and Begg [10] review the unique features of such Web 2.0 technologies, addresses questions regarding potential pitfalls, and suggests valuable applications in health education. The natural evolution of collaboration in these communities might depend on the ability to integrate applications developed by various organizations and individuals, offering relevant data and services customized for the specific community.

We argue in this paper that Web 2.0 *mashups* provides an adequate foundation for the development of systems addressing the above-mentioned needs. The next section presents the mashup concept and describes its advantages over previous offerings for the evolving needs of Health 2.0. The section after that presents a system called *Medic-kIT* and describes some of the main features it provides. We conclude by presenting the main challenges that will need to be addressed with the adoption of such technology.

2   The Proliferation of Web Mashups

Web Mashups is a new, fast-growing, integration approach in the data management field. Mashups aim to integrate not only data and services, but also live complex web-based applications. Although the theoretical basics of the concept are still being explored [11], there are already thousands of mashups available on the web. The website ProgrammableWeb.com [12] is an extensive repository listing thousands of mashups. Statistics provided by the site (see Figure 1) show a constant increase of approximately 100 mashups per month since September 2005.

Figure 2 shows the distribution of mashup classes. This distribution reflects the types of mashups and mashup APIs currently available on the web. (See, for example, the significant presence of mashups related to the "Mapping" domain, apparently due to the early exposure of APIs such as "Google Maps", "Yahoo! Maps", and others.) The distribution does not contain any significant presence of health-related mashups. We believe that the appearance of health-related mashup APIs in the near future will be followed by

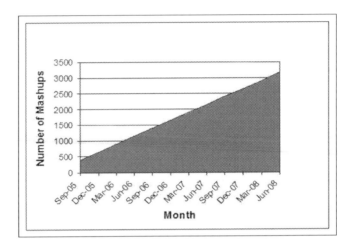

Fig. 1. Number of mashups - ProgrammableWeb website

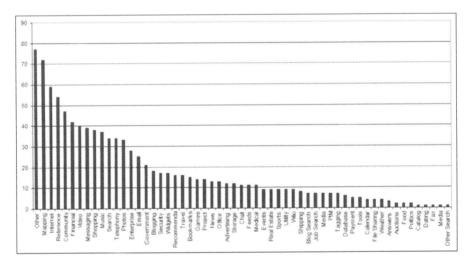

Fig. 2. Mashup class distribution - ProgrammableWeb website

the emergence of new health mashups, similar to the appearance of "Mapping" mashups that followed the exposure of "Mapping" APIs such as "Google Maps".

Mashups originated from the need to rapidly build situational applications that fit dynamic changes and needs. They are based on the understanding that the number of applications available on the web is growing rapidly. They quickly become complex, containing large amounts of heterogeneous data, varied functionalities, and built-in user interfaces. Therefore, in many cases, it would be an impossible mission for a single developer or even an IT department to build such applications in-house and integrate them with one another [13]. Based on standards, the goal of mashups is to provide an easy-to-use platform for users that integrates full heterogeneous applications for any kind of emerging use. Although they share common properties with portals, mashups differ in a few significant features [11] [14]. i) Mashups are designed to enable users to assemble customized complex applications, without requiring any programming skills; ii) The layout and functionality are dynamic and evolve over time; and iii) Mashups place a strong emphasis on high interoperability between the components, although they are heterogeneous from UI and functionality standpoints.

We suggest mashups as a leading technology for the requirements of Health 2.0. Platforms based on mashups have great potential to improve the quality of care and patient empowerment by delivering truly patient-centered, easy-to-use solutions; these can be easily customized for the needs of each specific patient, caregiver, and community. Mashups will enable the dynamic integration of heterogeneous data sources, applications, and services, and therefore can grow and evolve over time, together with the sharp increase in knowledge and services being offered on the web. They have the potential to enable various types of communication, coordination, and cooperation between all healthcare actors (e.g., patients, caregivers, and other service providers). Similar to our approach, Cho [15] suggests mashups as an appropriate technology for health librarians.

The rest of this section describes the detailed aspects of applying mashups to the healthcare domain.

2.1 Flexibility for the Long Tail

PHRs combine data, knowledge, and software tools, which help patients become active participants in their own care [4]. Our extension, the xPHR, contains additional components that help the patient address all personal needs, medical and others, in one platform. This includes components for medical data management (EMR), components enabling collaboration (e.g., chatting, media sharing), and others used for personal purposes (calendar, "To do list", maps). We suggest mashups as a convenient and flexible data management paradigm for xPHRs. Using the mashup approach, each user can easily select the set of components in which they are interested, and customize its layout, look, and feel according to their personal taste and habits. Our *Medic-kIT* system (see Fig. 3) provides such functionality for patients. Analysis carried out prior to the system design showed that, despite the various needs of patients, a particular set of components is essential for many medical scenarios. For example, a study performed on patient-couples undergoing IVF treatment[16] showed that 82% of them were interested in accessing their own medical data and 69% were interested in using it to communicate with their doctors. Similarly, patients enrolled in a web-based diabetes management program indicated that having access to the EMR and particularly the results of medical testing was important to them [17]. While these communities share common needs, each health condition has its specific requirements. For example, in the case of IVF treatment, a day planner component that displays an accurate timetable of treatment steps and predicts upcoming events may be essential for success [16], while patients with diabetes might benefit from the ability to upload blood glucose readings [17]. From an IT viewpoint, the mashup platform enables reuse of common components while being sufficiently flexible to suit the needs of the "Long Tail" of patients, as pointed out in other domains by Anderson [5].

2.2   Cooperative Patient-Centric Care

Patients are becoming more and more involved in their health treatments, closely monitoring the process, and mining and sharing information. They are willing to interact with multiple experts to acquire general knowledge about their condition and treatments, and obtain emotional support [18] and second opinions. For example, more than one-third of the visitors to an expert forum on involuntary childlessness sent detailed results of diagnostic tests and asked for a first or second opinion [18]. The growing use of patient-centric websites, such as [19] shows that in the near future, patients might want to manage their diseases from home, getting continuous remote monitoring services and support from their health providers. Ralston et al. [17] highlight the importance of such capabilities for diabetic patients. The internet has proved itself to be a simple, accessible, and attractive platform - where people feel comfortable sharing information and getting advice. Therefore, it seems natural that the Internet will serve as a basis for such a shift from passive patient participation in the healthcare process to a phase where patients become the treatment's center-of-gravity requires easy methods of communication, cooperation, and data exchange between patients, health providers, and other stakeholders. We claim that these goals can be addressed using mashups. The existence of collaboration mashup components (such as chatting, forums, blogs, or email) together with home-monitoring equipment and related software components will highly facilitate the data exchange and cooperation between patients and their caregivers.

Ash et al. [20] describe the kinds of silent errors that have been witnessed in healthcare systems and investigate the nature of these errors from social science perspectives (information science, sociology, and cognitive science). It has been found that these errors fall into two main categories: those occurring during the process of entering and retrieving information, and those occurring during the communication and coordination process that Patient Care Information Systems (PCIS) are supposed to support. The authors believe that by highlighting these aspects, PCISs can reduce the level of these silent errors. Enabling a seamless data flow between heterogeneous applications and data sources on the mashup platform (e.g., using manual drag-and-drop and automatic data flow) also assists in reducing medical errors [21]. For example, a direct link between a component that purchases medications, an EMR, and an SMS component, can ensure the timely consumption of medications prescribed by the doctor. Such mashup-based xPHRs can help send forms, pass on messages, and perform financial transactions. In this manner, xPHRs can reduce costs, save time, and improve the quality of the care given to patients.

Hardstone et al. [22] present the importance of informal discussions and provisional judgments as part of the collaborative process by which caregiver teams achieve consensual clinical management decisions over time. They argue that the greatest challenge for collaboration is to find ways of integrating such informality into traditional formal care record systems. Since there is a huge amount of available applications in the healthcare domain, we believe that the mashup approach, which has the ability to connect full applications, will be able to provide a customizable, flexible, component-based platform, that integrates medical applications with collaboration-oriented applications, as shown and described in [11].

2.3 Patients and Their Communities

Online support groups (OSGs) are communities connecting individuals who share a common problem. The communities related to health are mostly established by individuals with problems rather than by healthcare professionals. They involve mutual support as well as information provision [23]. As stated by citeVirtualCommunities, in April 2004, Yahoo! Groups listed almost 25,000 electronic support groups in the health and wellness section. Tuil et al. [16] shows that 55% of patients undergoing IVF treatment are interested in communicating with fellow patients. The proposed xPHRs can serve the community as a collaboration platform, in the same way that EMRs serve physicians. They will enable sharing of data, knowledge, and support among the members, via mashup components developed and maintained by the community. For example, a joint calendar could organize the community events and blogs. The community will need a back-end platform for data management that is sophisticated enough to handle both community-shared information and individuals' private data. An appropriate model for a suitable back-end might be the Independent Health Record Banks [24] vision, which enables an individual or any other entity to manage healthcare data and provide sets of services on top of it. Using such a platform, communities will no longer need to tie into a specific care provider, and will be able to suggest advanced care services to its members. This, in turn, will enable them to collaborate and share personal information with other community members.

3 *Medic-kIT*- System Description

To assess the potential of latent effectiveness in the mashup approach, we implemented *Medic-kIT*, a mashup-based xPHR system. A screenshot of the system is displayed in Figure 3. The system follows the objectives of xPHRs and the requirements presented in [16], and includes components of three main classes: i) Medical, ii) Personal, and iii) Collaboration. The first class contains components such as the PHR and RSS feeds related to the medical domain. The second contains "My notes", a "To do list", an image viewer (for scanned documents and figures), a map, and a calendar. The third contains components such as SMS, email, and blogs. The application was presented at the 2008 Mobile World Conference (3GSM), as part of a larger system that provides personalized monitoring of patients with notification on anomalies to relatives and caregivers. The front-end of our system is implemented using Adobe ® Flex ™2, Rich Internet Application development tool set. The back-end was designed to handle both the system's internal data (e.g., layout and component configuration), and content collected and managed by the system (e.g., RSS feeds and patient details). The front-end is based on an IBM ® DB2 ™ v9 database management system that provides the storage and query of XML documents. The data access infrastructure is wrapped in a REST+ATOM interface layer, providing capabilities and communication formats aligned with Web 2.0 principles. On top are two additional layers: *i)* Event-Manager, which handles the events dispatched as a result of interaction with the user or between components, and *ii)* Transformation-Manager, which enables transformation of data being passed between the system components. The easy-to-use drag-and-drop feature enabled by our system

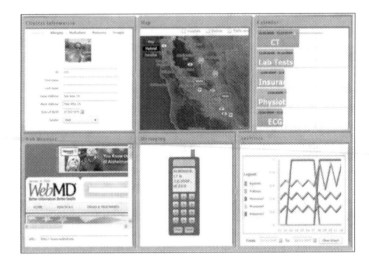

Fig. 3. *Medic-kIT* - a mashup-based xPHR system

demonstrates the importance of these two layers. The interaction between the two layers contributes to a more intuitive and compelling usage experience, reduces errors [21], and therefore improves treatment quality.

4   Conclusions and Research Opportunities

Individuals in today's world are gaining more control over all aspects of their life. In particular, healthcare patients are becoming more aware and involved in their own care process. They understand their medical condition better and do not rely entirely on a single physician for all their health needs. The IT world encourages this trend by increasing exposure to great variety of healthcare knowledge domains, information databases, and services. These opportunities increase the demand for supporting tools and software components. In this paper we proposed mashups as a leading technology for facing the emergent needs in the Health 2.0 era. We discussed related aspects and demonstrated how mashups could assist in coping with them.

Although the inherent advantages of the mashup approach and its potential for the healthcare domain are clear, it is still a technology under exploration. Further study might reveal additional research questions in several areas in academia and industry.

- *Mashup composition and design*: The mashup approach empowers users with the ability to design their applications according to changing needs. Given a large number of applications and additional knowledge about the ways they are connected, users might want to benefit from this cumulative experience for their specific requirements. Abiteboul et al. [25] demonstrates a system that assists users in composing their personal mashups, based on their needs and on the experiences of other users. Riabov et al. [26] dealt with a similar question on data mashups.

- *Privacy and security:* Web 2.0 technologies raise various challenges related to security [27]. Healthcare data must be highly secure, especially when maintained on the Internet and shared with other individuals, mashup components, and services. This important issue will require the development of secure mechanisms for communication, as proposed by [28,29,30,31]. Due to the variety of roles involved in every scenario (e.g., patient, doctor, insurer, hospital), each with its own requirements, role-based access control models will have to be considered and implemented [32].

- *Quality assurance:* The quality of information and knowledge in the health domain is acute [33], since misleading data could lead to life-threatening results. Therefore, mechanisms controlling the quality of data and services being offered must be developed for mashups [34].

- *Flow and process management:* Processes in the health domain are usually complex. Thus, systems that enable monitoring, navigation, and intervention in the flow will be required [35].

In our view, these directions should be addressed in order to make such a xPHR system valuable and effective for entities in the healthcare domain.

References

1. Rheingold, H.: The Virtual Community: Homesteading on the Electronic Frontier, revised edition. The MIT Press, Cambridge (2000)
2. Health 2.0 conference, http://www.health2con.com/
3. Pratt, W., et al.: Incorporating ideas from computer-supported cooperative work. J. of Biomedical Informatics 37(2), 128–137 (2004)
4. Tang, P., et al.: Personal Health Records: Definitions, Benefits, and Strategies for Overcoming Barriers to Adoption. J. Am. Med. Inform. Assoc. 13(2), 121–126 (2006)
5. Anderson, C.: Wired 12.10: The long tail
6. Sittig, D., King, S., BL, H.: A survey of patient-provider e-mail communication: what do patients think? Int. J. Med. Inform. 61(1), 71–80 (2001)
7. Angst, C.M., Agarwal, R.: Patients Take Control: Individual Empowerment with Personal Health Records. SSRN eLibrary (2004)
8. Beenen, G., et al.: Using social psychology to motivate contributions to online communities. In: CSCW, pp. 212–221. ACM, New York (2004)
9. Eriksson-Backa, K.: Who uses the web as a health information source? Health Informatics J. 9(2), 93–101 (2003)
10. McGee, J.B., Begg, M.: What medical educators need to know about "web 2.0". Medical teacher 30(2), 164–169 (2008)
11. Abiteboul, S., Greenshpan, O., Milo, T.: Modeling the mashup space. In: Chan, C.Y., Polyzotis, N. (eds.) WIDM, pp. 87–94. ACM, New York (2008)
12. Programmableweb, http://www.programmable.com/
13. Jhingran, A.: Enterprise information mashups: Integrating information, simply. In: VLDB, pp. 3–4. ACM, New York (2006)
14. Daniel, F., et al.: Understanding ui integration: A survey of problems, technologies, and opportunities. IEEE Internet Computing 11(3), 59–66 (2007)
15. Cho, A.: An introduction to mashups for health librarians. JCHLA/JABSC 28, 19–22 (2007)

16. Tuil, W., et al.: Patient-centred care: using online personal medical records in IVF practice. Hum. Reprod. 21(11), 2955–2959 (2006)
17. Ralston, J., et al.: Patients' experience with a diabetes support programme based on an interactive electronic medical record: qualitative study. BMJ 328(7449), 1159 (2004)
18. Himmel, W., et al.: Information needs and visitors' experience of an internet expert forum on infertility. J. Med. Internet Res. 7(2), e20 (2005)
19. Patientslikeme, http://www.patientslikeme.com/
20. Ash, J.S., Berg, M., Coiera, E.: Some unintended consequences of information technology in health care: The nature of patient care information system-related errors. Journal of the American Medical Informatics Association 11(2), 104–112 (2003)
21. Ash, J., et al.: Some unintended consequences of information technology in health care: The nature of patient care information system-related errors. J. Am. Med. Inform. Assoc. 11(2), 104–112 (2004)
22. Hardstone, G., et al.: Supporting informality: team working and integrated care records. In: Proc. CSCW, pp. 142–151. ACM Press, New York (2004)
23. Potts, H.: Online support groups: an overlooked resource for patients. He@lth Information on the Internet 44(3), 6–8 (2005)
24. Shabo, A.: A global socio-economic-medico-legal model for the sustainability of longitudinal electronic health records, part 1. Methods of Information in Medicine 45(3), 240–245 (2006)
25. Abiteboul, S., Greenshpan, O., Milo, T., Polyzotis, N.: Matchup: Autocompletion for mashups. In: ICDE (2009)
26. Riabov, A.V., Boillet, E., Feblowitz, M.D., Liu, Z., Ranganathan, A.: Wishful search: interactive composition of data mashups. In: WWW 2008: Proceeding of the 17th international conference on World Wide Web, pp. 775–784. ACM, New York (2008)
27. Lawton, G.: Web 2.0 creates security challenges. Computer 40(10), 13–16 (2007)
28. Jackson, C., Wang, H.J.: Subspace: secure cross-domain communication for web mashups. In: WWW 2007: Proceedings of the 16th international conference on World Wide Web, pp. 611–620. ACM, New York (2007)
29. Erlingsson, Ú., Livshits, B., Xie, Y.: End-to-end web application security. In: HOTOS 2007: Proceedings of the 11th USENIX workshop on Hot topics in operating systems, Berkeley, CA, USA, USENIX Association, pp. 1–6 (2007)
30. Davidson, M., Yoran, E.: Enterprise security for web 2.0. Computer 40(11), 117–119 (2007)
31. De Keukelaere, F., Bhola, S., Steiner, M., Chari, S., Yoshihama, S.: Smash: secure component model for cross-domain mashups on unmodified browsers. In: WWW 2008: Proceeding of the 17th international conference on World Wide Web, pp. 535–544. ACM, New York (2008)
32. Barkley, J.: Comparing simple role based access control models and access control lists. In: RBAC 1997: Proceedings of the second ACM workshop on Role-based access control, pp. 127–132. ACM, New York (1997)
33. Pawlson, L.G.: An introduction to quality assurance in health care - avedis donabedian, p. 240. Oxford university press, New York (2003); $37.95, hardcover. isbn 0-19-515809-1; American Journal of Preventive Medicine 26(1), 96 (2004)
34. Zahoor, E., Perrin, O., Godart, C.: Mashup model and verification using mashup processing network (2008)
35. Anyanwu, K., Sheth, A., Cardoso, J., Miller, J., Kochut, K.: Healthcare enterprise process development and integration. Journal of Research and Practice in Information Technology, Special Issue in Health Knowledge Management, 83–98 (1999)

Semantic Warehousing of Diverse Biomedical Information

Stefano Bianchi[2], Anna Burla[1], Costanza Conti[3], Ariel Farkash[1], Carmel Kent[1],
Yonatan Maman[1], and Amnon Shabo[1]

[1] IBM Haifa Research Labs, Haifa University, Mount Carmel, 31905, Haifa, Israel
[2] Softeco Sismat S.p.A., Via De Marini 1, WTC Tower, 16149, Genoa, Italy
[3] IMS-Istituto di Management Sanitario SRL - via Podgora, 7-20122 Milano, Italy

Abstract. One of the main challenges of data warehousing within biomedical
information infrastructures is to enable semantic interoperability between its
various stakeholders as well as other interested parties. Promoting the adoption
of worldwide accepted information standards along with common controlled
terminologies is the right path to achieve that. The HL7 v3 Reference Information Model (RIM) is used to derive consistent health information standards such
as laboratory, clinical health record data, problem- and goal-oriented care, public
health and clinical research. In this paper we describe a RIM-based warehouse
which provides (1) the means for data integration gathered from disparate and
diverse data sources, (2) a mixture of XML and relational schemas and (3) a
uniform abstract access and query capabilities serving both healthcare and
clinical research users. Through the use of constrained standards (templates),,we
facilitate semantic interoperability which would be harder to achieve if we only
used generic standards in use cases with unique requirements. Such semantic
warehousing also lays the groundwork for harmonized representations of data,
information and knowledge, and thus enables a single infrastructure to serve
analysis tools, decision support applications, clinical data exchange, and
point-of-care applications. In addition, we describe the implementation of that
semantic warehousing within Hypergenes, a European Commission funded project focused on Essential Hypertension, to illustrate the unique concepts and
capabilities of our warehouse.

Keywords: Biomedical, HL7v3, RIM, CDA, Repository, XML, Query, EAV,
Data Warehouse.

1 Introduction

Healthcare information systems typically contain data and knowledge related to a
specific health domain with idiosyncratic semantics [1] . As such they constitute silos
of information that are hard to integrate [2]. Health information warehouses are established in an attempt to accomplish such integration and support patient-centric care [3]
as well as secondary use of the data such as analysis of aggregated data in the context of
clinical research, quality assurance, operational systems optimization, patient safety

Y.A. Feldman, D. Kraft, and T. Kuflik (Eds.): NGITS 2009, LNCS 5831, pp. 73–85, 2009.

and public health [4]. In healthcare, the emerging concept of personalized care involves taking into account the clinical, environmental and genetic makeup of the individual [5, 6]. To that end, a warehouse may serve as a main information source for putting together the patient electronic health record (EHR) where the consistent and explicit semantics is crucial for machines to reason about the record [7]. Thus, there is a tension between extensive warehousing of data with diverse semantics against efforts that need coherent semantics such as evolving a longitudinal EHR or performing a more focused analysis based on deep understanding the data. This paper describes a solution to this fundamental problem by proposing an approach of semantic warehousing based on a low level information model serving as a common language to represent health data and knowledge. To that end, the HL7 v3 Reference Information Model [8] (RIM) is used to derive a specific data model for the data warehouse – a core component of a Biomedical Information Infrastructure (BII).

1.1 The Reference Information Model

The RIM has been developed and approved in the past decade by the international community of standards developing organizations and is now an ANSI as well as ISO-approved standard. The RIM provides a unified 'language' to represent actions made by entities. Health data is described by associations between entities who play roles that participate in actions. For example, an organization entity plays a role of laboratory that participates in an observation action, or, a person entity plays a role of a surgeon who participates in a procedure action, and so forth. Actions are related to each other through "act relationship" providing the mechanism to describe complex actions. The RIM includes the unique attributes of each of the entities, roles, participants and actions that are relevant to health, described in an object-oriented manner, e.g., the observation action has a value attribute while the procedure action has a target site attribute, etc.

The RIM is then used by the various standardization groups to develop domain-specific standards such as laboratory, pharmacy and clinical documents to name just a few. Typically, those domains are horizontal and generic in nature and can be used across the various clinical specialties. All specs have XML implementations (W3C Schemas) to easily capture the complex semantics.

1.2 XML-Based Warehousing

RIM-based applications and repositories have been developed thus far based on relational schemas [9, 10]. Our approach has been to use the XML representations of the specifications as well as the data instances to natively handle and persist the data in an XML-enabled database system. The access to the data and knowledge persisted in our warehouse is challenging because of the mixture of data types (clinical, environmental and genetic data), mixture of persistency schemas (XML, relational and binary data) but most importantly due to the mixture of different access patterns in healthcare versus research. In healthcare, the access is typically per patient and seeks whole instances (e.g., complete discharge summaries and operative notes of a patient at the point of care) while in research the access is to different grouping of discrete data based on complex criteria and interdependencies, typically not patient-centric rather centered around diseases, medications, etc. These differences pose a challenge in designing a data access layer that is effective to all use cases. While data entry is based on convergence and strives to insert merely standardized data, the data access layer described in this paper facilitates various access methods to accommodate different users and use cases and thus spans from direct use of XQuery and SQL for 'super-users' to APIs used by domain IT experts to front-end applications used by end-users. A central mechanism to achieve this variety of data access options is by indexing and promoting XML structures to an abstract table, e.g., in the form of Entity-Attribute-Value tuples where each tuple represents the result of some semantic computation on XML structures in the instances stored in the warehouse. This mechanism is described in more detail in the data access section of this paper.

The following sections describe the main two layers surrounding our warehouse, i.e., the Data Entry and Data Access layers. These two layers along with the persistency layer constitute the warehouse component which is the core of a Biomedical Information Infrastructure.

2 Data Entry Process

Inserting data into the warehouse, especially when dealing with rich and diverse data from disparate data sources, is a complicated task. The data needs to be harmonized, validated, normalized, and integrated into common structures that can be accommodated by the warehouse. In addition, relationships among data items are often defined implicitly and should be expressed in an explicit and standard way so that analysis algorithms not aware of the implicit semantics could run effectively. Finally, data should be persisted in a way that will enable semantic computation and interoperability, thus laying the grounds for effective query and retrieval against the warehouse.

2.1 Harmonization

Integration of data coming from dissimilar data sources about the same clinical, environmental and genetic phenomena must first undergo a process of conceptual harmonization, i.e. convergence of the sources metadata to a single and agreed-upon terminology. Take for example blood pressure measurement variables from three different cohorts of essential hypertension (see figure 1).

In order to be able to compare data of different cohorts, one should first map all cohort variables to a core terminology. The core terminology is delegated to a standard terminology such as LOINC, SNOMED CT and ICD. We used the Web Ontology Language [11] (OWL) to map all cohort variables to a core ontology. The process starts by creating a cohort class (OWL class) for each variable and then mapping it to a core ontology class by specifying an equivalent class relationship. In case where additional parameters are needed to refine the cohort variable definition, a cohort instance (OWL individual) is created with the needed data parameters (e.g., temporal parameters, see figure 2).

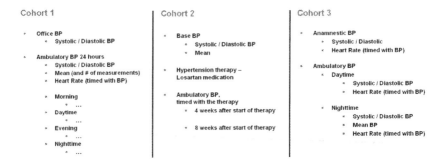

Fig. 1. Similarity and disparity in blood pressure measurement schemes

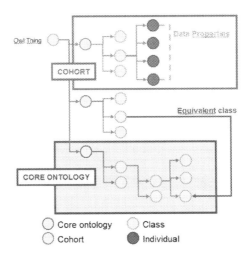

Fig. 2. Ontology schematic diagram

Thus, following the example depicted in figure 1, Cohort 2 Ambulatory Blood Pressure would need two individuals that would have the extra specification for 4 or 8 weeks after therapy. Once metadata is harmonized and standardized, data received from data sources should first be verified that it is adequate for its corresponding metadata under the given terminology. The data should then be normalized, e.g., different value sets with the same or similar semantics like yes/no versus true/false. Finally the proprietary data is ready to be transformed to a standard instance.

2.2 Capturing Richness of Data

Having similar sets of metadata represented in an agreed-upon terminology provides the basis for semantic interoperability [12], i.e. the ability to compare two orphan items of data and reason about its level of similarity. However, biomedical data is typically complex, consisting of associations and dependencies among discrete data items as well as among common structures. As aforementioned, in the HL7 Reference Information Model (RIM), actions are related to each other through "act relationships" providing the mechanism to describe complex actions. Let's look once more at the example in figure 1: In Cohort 2, the Ambulatory Blood Pressure measurement is measured while the subject is treated by a medication for hypertension called Losartan. This calls for the association of the act of observing the blood pressure to the act of administering the drug so that semantics if explicitly represented in the warehouse. This information is sometimes crucial for the physician; high blood pressure measurement while under Losartan regimen has a completely different meaning then without such intervention. Another example would be when Blood Pressure and Heart Rate measurements are timed with a diagnostic procedure such as Echocardiography. In this case, Blood Pressure and Heart Rate values are not relevant for the diagnosis of follow up of hypertension itself, since measured when the patient is in a potential stressful situation, but are significant in interpreting Echocardiography findings.

Therefore, to capture the full richness of the data those associations should be established, preferably during the data entry process when the experts responsible for the data source can provide the implicit semantics often hidden in unstructured documentation or merely in their minds. To that end, the HL7 v3 Reference Information Model (RIM) is used to derive a specific data model for the data warehouse.

As aforementioned, the RIM provides a unified 'language' to represent health actions such as observations, procedures and substance administrations. It facilitates the explicit representations of the rich semantics of the data. In the examples discussed above, the blood pressure and heart rate measurements are represented as RIM observations and when appropriate, these observations are associated with an echocardiography procedure or with a substance administration of Losartan.

2.3 Data Persistency

The HL7 Reference Information Model (RIM) is only the underlying meta-model for expressing domain-specific data. As aforementioned, various standardization groups develop generic standards such as laboratory, pharmacy and clinical documents. Those RIM-derived standards have XML implementations (W3C XML Schemas) to capture the complex semantics in the dominant structure for content representation today [13]. XML instances that follow these schemas are adjusted to the jargon of the domain. For example a Clinical Document Architecture (CDA) instance will use terms such as: section, entry and consumable, whereas these terms are mapped to Act [classCode=DOCSECT], ActRelationship [typeCode=COMP], and Participation [typeCode=CSM] correspondingly, under the RIM meta-model. In order to be able to manipulate, process, query and retrieve instances from all domains in the data warehouse. We apply a process that we tagged as 'RIMification' whereby we apply a transformation from the XML instance of the domain to an XML instance that follows the XML schema generated from the RIM UML. We then take the "RIM instance" and use IBM DB2 9.5 pureXML capabilities [14] to natively persist it as XML. Having instances from all domains represented as instances of the same underlying model, and having them in their original XML structure, enables us to build a more generic API for querying the warehouse in a more consistent way. Furthermore, the above method facilitates retrieval of an original instance, or parts of the instance, by using the reverse process of RIMification so that it is seamless to the end user. This API and related tools will be described in detail in a later section.

2.4 Semantic Interoperability

When establishing a new warehouse, it is important to select the most appropriate RIM-based standards and how they would relate to each other, as well as what constraining could be applied to those generic standards. These efforts lead to the warehouse data model which is developed in a two-step constraining process:

1. **Domain Standards selection:** Existing standards are selected to be the basis of the data model, for example, the standards Clinical Document Architecture (CDA); Clinical Genomics - Genetic Variation (GV) and Family History (FH) could be a good basis for a clinical genomics research solution involving population genetics.
2. **Templates development:** A template is an expression of a set of constraints on the RIM or a RIM derived model that is used to apply additional constraints to a portion of an instance of data which is expressed in terms of some other Static Model. Templates are used to further define and refine these existing models to specify a narrower and more focused scope [15]. The standards selected for the warehouse data model are further constrained to create templates of the standards which reflect the specific structure required by the customer, e.g., the precise way to represent blood pressure measurements in different settings such as ambulatory during the night. Note that this step includes merely constraining of the selected standards so that instances are always valid against the generic standard but also comply with the templates [16].

As stated above, we use the RIM in order to capture the richness of the data. Nevertheless, richness lead sometimes to diversity and thus we use templates as a means to facilitate semantic interoperability among interested parties by narrowing down the large number of compositional expressions allowed by the RIM to a nailed-down structure for each piece of data. By supplying a "closed template", i.e. a strict template specification constraining each datum to a specific location within the XML, we enable users and utilities, such as decision support tools, unequivocal access to the biomedical data stored in the warehouse.

3 Data Access

The need for effective query capabilities on top of various RIM-based data instances created from diverse data sources is the main challenge in developing the data access layer. To cope with this challenge, we developed a promotion mechanism bridging between XML complex data structures and data consumers in various scenarios (e.g., clinical care, public health and clinical research applications, analysis tools etc.).

RIM-based specifications are translated to W3C XSD schemas reflecting different levels of complexities in various parts of a specification. For example, CDA header, which defines the document content metadata (e.g., patient ID, document creation time, document author etc.), is less permissive for recursive structures, optional elements and so forth than the CDA body. Searching for information in the CDA header, is very straightforward using XPATH or other XML processing techniques. On the other hand, other parts in the CDA allows for recursive structures and are much less constrained elements in order to allow diverse usage and styles of the CDA from different use cases and applications. The same clinical observation might be represented in two CDA documents by completely different structures, using a different set of terminologies for assigning codes to attributes of those structures. This flexibility allows for high expressibility while limits the interoperability of the data. The same XML processing techniques which were sufficient for searching the CDA header will not suffice in these cases; XML schemas on top of RIM can be used for the purpose of XML validation but cannot solve the challenge of fully understanding the semantics of a data structure.

3.1 The Promotion Layer

The Promotion Layer is used to promote specific user predefined XML elements/structures and allow for an abstract and direct way for the data consumer to access the data (see Figure 3). This mechanism decouples the data consumer from the need to be very familiar with the specification XML structure and allows the consumer to concentrate on the business logic of interest. For this purpose we implemented the Promotion Layer by the means of an Extended Entity-Attribute-Value (EAV model), which is a well accepted data model in the clinical data warehousing field [17, 18, 19]. In the EAV data model, each row of a table corresponds to an EAV triplet: an entity, an attribute, and an attribute's value. The Entity of our EAV corresponds to the instance id. For example, the entity "CDA document no. 111" has the attribute "PatientID" with the value of "2.16.840.1.113883.3.933". Similarly, the same entity can also have the attribute "Diastolic Blood Pressure" with an integer value of 100.

The promoter (who is typically not an end user) chooses which information elements from the XML data to promote for access into the EAV table, while optionally performing semantic computations on the promoted attributes (e.g., context and terminology computations). The main drawback when using the classic EAV model is the loss of the context, which in our warehouse, resides in the hierarchical structure of the XML. To that end, we have extended the basic EAV model to allow for regaining the depth of the XML data, for example, when the user needs to see a broader context of retrieved data.

We are supporting three levels of semantic computations:

1. **Scope:** For each promoted attribute, we allow for back referencing to the XML element, which originated the attribute. The promoter is able to choose the scope of the back referenced XML snippet. For example, the immediate containing CDA Observation structure; the whole CDA Section structure; etc. The data consumer can retrieve the containing XML snippet and derive the context of promoted attribute at stake. Getting the containing XML snippet will inform the Data Consumer with additional information on the promoted attribute. For example, sibling attributes like descriptions in free text, status, additional codes and patient id as well as associations to other structures in cases when the structure is part of a broader clinical statement.

2. **RIM-based computations:** There are cases, in which the required context does not reside simply in the contained XML snippet, and its computation involves traversal of other parts of the XML and calculating more complex XML paths. For

this purpose, we provide services to conduct context for a specific calculated location in the data instance. For example, once a specific observation in CDA was promoted, there may be a need to get the identification of the associated patient or subject. The naïve way is retrieving the patient associated with the CDA document (expressed by the Entity of the EAV record), which reflects the top-down propagation nature of XML based instances. Other subjects may be indicated in various locations in the document overriding the mandatory subject specified in the document header. Thus, there is a need for a more sophisticated calculation in order to retrieve the accurate value, and this computation is offered to the promoter as part of the promotion API. Another example could be the calculation of the computation of RIM's negation indicators as seen in the example in figure 3. Temporal calculations are needed as well because the time data types of the RIM are complex to accommodate multipart prescription for example.

3. **Clinical computations:** Other semantic computations, for example, a clinical computation of the average of a few blood pressure examinations taken prior to a specific therapy, could be promoted instead of a single examination value, while making the average value a better basis for comparison to other values than a single measurement, as defined in the CDA document.

While there isn't yet an agreed-upon formalism for representing constraints against RIM-based specifications (i.e., Templates, see Data Entry section), the Promoter should be familiar with the queried XML structures. When such constraining formalism is agreed, the Promoter will be able to utilize it for promoting XML elements and attributes by using templates without having to be familiar with the RIM structure of the instance/document.

3.2 Tools for Data Access

The Promotion Layer both simplifies and abstracts the access to the data promoted to a flat, generic and relational schema while preserving its rich semantics through the back referencing to the source XML. We realize that one of the main challenges is to allow for access tailored to the requirements of different use cases from both healthcare and clinical research domains. To this end, we designed two Data Access tools, both built on top of the Promotion Layer, for different types of Data Consumers who are not accustomed or skilled in direct XQUERY access to the XML data (see figure 3).

3.3 Java Access API

The Java Access API on top of the Promotion Layer allows for Application Data Consumers to query information promoted by the Promotion Layer beforehand. The API is constructed by two layers: The Query Layer queries the data and returns full EAV records while the Retrieval Layer retrieves the containing XML snippet given an EAV record. The Query Layer has a notion of a simple SQL SELECT statement. It consists of a set of promoted attributes to be part of the result space (SELECT Clause) and a set of logical conditions on the result (WHERE Clause). The Query Layer assumes an implicit join between attributes in the SELECT Clause and in the WHERE Clause based on the Entity ID. Using the Query Layer enables the Data Consumer to execute a query and receive the relevant EAV records (a set of Entities, Values, and

Back-References). In the case where more contextual information is needed, the Retrieval layer can extract the containing XML snippet for further processing. The Context Conduction services can be used on top of the Java Access API, in order to enrich the computational capabilities.

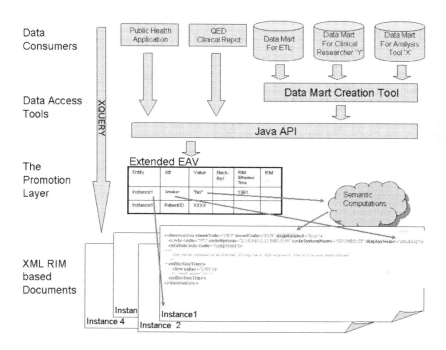

Fig. 3. Data Access Architecture: The promotion Layer, facilitate the picking of a subset of attributes from the underlying warehouse, a set of predefined semantic computations performed upon this set of promoted data and the adaptation of it to the data-consumer view. This mechanism lays the ground for data mart creation tool, and application using the java API in order to approach and consume the promoted data.

3.4 Datamart Creation Tool

XML schema model has its advantages in the biomedical warehousing field. Nevertheless, we recognize that most of the existing analysis tools for clinical research have interfaces to relational models and schemas to get the data for the analysis. A Data Mart is a relational, user- requirements oriented logical subset, or view of the full warehouse. Besides the fact that Data Marts give a relational facade for XML based data (and by that allow traditional SQL-based access), they also have the benefit of easy access to frequently needed data that has been pre-processed and the option to design a schema oriented to the Data Consumer requirements rather than to the Data Source schema. As a result, Data Marts may improve the end-user response time, and simplicity. The Data Mart Creation tool uses the Promotion Layer's infrastructure for administrating the schema of Data Marts. A set of promoted attributes serve as the specification for the

Data Mart schema, then, through the use of the Data Access API (discussed above) the Data Mart is derived from a selected set of promoted attributes. It can be persisted as a relational table, or just exported as a CSV file.

4 Use Cases

4.1 Hypergenes

Hypergenes is a European Commission funded project that aims at building a method to dissect complex genetic traits using essential hypertension as a disease model [20]. Most of the common-complex, chronic diseases, that have a high prevalence in our populations, arise through interaction between genetic, environmental and life-style factors. To understand the composite origin of these diseases, there is a need to know the path from genotype to phenotype. To that effect, the Hypergenes consortium includes numerous data sources for genomic, clinical and environmental data. This scenario was an ideal use case for our data warehouse since it featured an array of capabilities among which were the following: data entry of genomic, clinical and environment data from disparate data sources, and data access by a versatile set of consumers ranging in usage from research to healthcare.

The first phase included collection and harmonization of the variables of 12 cohorts and mapping the cohort variables to core ontology. The core ontology was built using a mixed top-down and bottom up approach, in order to simplify the overview of information to be gathered and to define a simple taxonomy of concepts to be used as a reference mapping for the variables provided by the clinical partners. OWL ontology was built to represent the aforementioned mapping. Data extracted from the cohorts was analyzed; the process involved normalization of proprietary values such as internal enumerations, e.g. 0=Male, 1=Female, etc.

The second phase was to build corresponding CDA documents and Genetic Variation instances (see section on the warehouse data model) with standard terminology for the given data. Cohort variables meaning and semantic relations were elucidated by the cohort scientific leaders and expressed explicitly. Essential hypertension templates were constructed in order to capture rich data while laying the grounds for interoperable data access. All XML instances were then 'RIMified' and stored in the data warehouse.

Hypergenes assembles data analysis layer, done by machine learning experts. This layer laid the need for the warehouse data to be accessible to data mining, analysis and research tools, while also requires the warehouse to be able to re-store and integrate the information and knowledge resulting from those analysis tools. The Hypergenes use case's special flavor comes from the enablement of a warehouse based on data model, which is more healthcare in nature, to the special IT needs of the analysis or clinical research domains. For example, HL7 RIM-based XML instances and documents are designed for patient centric data exchange within a healthcare scenario, while analysis tools also require relational, mass data, optionally de-identified, and flexible schema construction. For this purpose, our Data Mart Creation tool, leaning on the Promotion Layer, facilitate these user-tailored transformations.

4.2 Query for Existing Data

IHE [12] (Integrating the Healthcare Enterprise) is an initiative by healthcare professionals and industry to improve the way computer systems in healthcare share and exchange information. It defines Integration Profiles which describe clinical information management & interoperability use cases and specify how to use existing standards (e.g., HL7, DICOM) to address these use cases.

One of the Integration profiles is Query for Existing Data [21] (QED), which supports dynamic queries for clinical data. It defines a use-case in which a Clinical Consumer System queries a Clinical Source System for clinical information. The warehouse implements a QED Clinical Source System, as an integral part of it.

QED defines different types of queries, which are all patient centric. A Clinical Consumer System can ask for various types of clinical data for a specific patient, for example: vital signs, allergies, medications, immunizations, diagnostic results, procedures and visit history. The profile leverages HL7 domain standards for the content model and the common HL7 messaging formats for conveying both query and result. The response is composed by CDA snippets that describe clinical data according to the query's parameters. Since the warehouse's data model is based on the HL7 RIM, the implementation of a Clinical Data Source System on top of our warehouse Data Access API has been a relatively straightforward task.

Having a data warehouse that supports Clinical Data Source System, helps healthcare organizations to increase quality of treatment by getting access to the most relevant and timely core health information about a patient and by that promote optimal patient health care. Natural continuation of QED effort can be creating a service that generates a CCD [21] (Continuity of Care Document) for a patient out of a set of queries on the Clinical Data Source System.

5   Conclusions

We have shown the feasibility of having semantic warehousing, that is, how it is possible to have on the one hand a warehouse that accommodates a variety of health data types, structures, and semantics sent from different sources while having all of this diversity represented with a common underlying language that can express rich semantics. In this way the warehouse is capable of manifesting the similarities in the incoming data while keeping its disparities. The underlying language is reflected in the persistency layer of the warehouse as well as in the unique processes of data entry and retrieval. An important feature of using this underlying language is the ability to create domain models and further constrain to any level of specificity. This gives rise to performing analysis over the data warehouse without intimate knowledge regarding the idiosyncratic structure of each data source, e.g., data mining that looks for adverse drug events in a large number of medical records from many sources. At the same time, if certain data sets in the warehouse have been further constrained, then it is possible to perform more focused analysis or participate in information exchange accomplishing semantic interoperability. Furthermore, constrained data sets in the ware house could be exported to more accessible formats like tabulated data marts and we have described mechanisms to facilitate this export process in an automated way.

Acknowledgments. The research leading to these results has received funding from the European Community's Seventh Framework Program FP7/2007-2013 under grant agreement n° 201550.

References

1. Kalra, D., Lewalle, P., Rector, A., Rodrigues, J.M., Stroetmann, K.A., Surjan, G., Ustun, B., Virtanen, M., Zanstra, P.E.: Semantic Interoperability for Better Health and Safer Healthcare. In: Stroetmann, V.N. (ed.) Research and Deployment Roadmap for Europe. SemanticHEALTH Project Report (January 2009), Published by the European Commission, http://ec.europa.eu/information_society/ehealth
2. Knaup, P., Mludek, V., Wiedemann, T., Bauer, J., Haux, R., Kim, L., et al.: Integrating specialized application systems into hospital information systems – obstacles and factors for success. Stud. Health Technol. Inform. 77, 890–894 (2000)
3. Gold, J.D., Ball, M.J.: The Health Record Banking imperative: A conceptual model. IBM Systems Journal 46(1) (2007)
4. Bock, B.J., Dolan, C.T., Miller, G.C., Fitter, W.F.: The Data Warehouse as a Foundation for Population-Based Reference Intervals. American Journal of Clinical Pathology 120, 662–670 (2003)
5. Ruano, G.: Quo vadis personalized medicine? Personal Med. 1(1), 1–7 (2004)
6. Davis, R.L., Khoury, M.J.: The journey to personalized medicine. Personal Med. 2(1), 1–4 (2005)
7. Protti, D.: Moving toward a Single Comprehensive Electronic Health Record for Every Citizen in Andalucía, Spain. Electronic Healthcare 6(2), 114–123 (2007)
8. HL7 Reference Information Model, Health Level Seven, Inc., http://www.hl7.org/v3ballot/html/infrastructure/rim/rim.htm
9. Eggebraaten, T.J., Tenner, J.W., Dubbels, J.C.: A health-care data model based on the HL7 Reference Information Model – References. IBM Systems Journal, Information-Based Medicine 46(1) (2007)
10. Spronk, R., Kramer, E.: The RIMBAA Technology Matrix. A white paper published at, http://www.ringholm.de/docs/03100_en.htm
11. Web Ontology Language, http://www.w3.org/TR/owl-features/
12. Heiler, S.: Semantic interoperability. ACM Computing Surveys 27(2), 271–273 (1995)
13. Shabo, A., Rabinovici-Cohen, S., Vortman, P.: Revolutionary impact of XML on biomedical information interoperability. IBM Systems Journal 45(2) (2006), http://www.research.ibm.com/journal/sj/452/shabo.html
14. IBM DB2 pureXML, http://www.ibm.com/db2/xml
15. Template project on the HL7 ballot site, http://www.hl7.org/v3ballot/html/infrastructure/templates/templates.htm
16. Li, J., Lincoln, M.J.: Model-driven CDA Clinical Document Development Framework. In: AMIA Annu. Symp. Proc. 2007, October 11, p. 1031 (2007)
17. Nadkarni, P.M., Brandt, C., Frawley, S., Sayward, F.G., Einbinder, R., Zelterman, D., Schacter, L., Miller, P.L.: Managing Attribute-Value Clinical Trials Data Using the ACT/DB Client-Server Database System. Journal of the American Medical Informatics Association 5(2), 139–151 (1998)
18. Nadkarni, P.: The EAV/CR Model of Data Representation, http://ycmi.med.yale.edu/nadkarni/eav_cr_frame.htm

19. Wang, P., Pryor, T.A., Narus, S., Hardman, R., Deavila, M.: The Web-enabled IHC enterprise data warehouse for clinical process improvement and outcomes. In: Proceedings AMIA Annu. Fall Symp. (1997)
20. Hypergenes, An FP7 European Commission Funded Project:
    http://www.Hypergenes.eu/
21. QED (Query for Existing Data) profile,
    http://wiki.ihe.net/
    index.php?title=PCC-1#Query_Existing_Data
22. HL7 CCD (Continuity of Care Document),
    http://www.hl7.org/documentcenter/public/
    pressreleases/20070212.pdf
23. Shabo, D.D.: The seventh layer of the clinical-genomics information infrastructure. IBM Systems Journal 46(1) (2007)
24. Shabo, A.: Integrating genomics into clinical practice: standards and regulatory challenges. Current Opinion in Molecular Therapeutics 10(3), 267–272 (2008)
25. Shabo, A.: Independent Health Record Banks – Integrating Clinical and Genomic Data into Patient-Centric Longitudinal and Cross-Institutional Health Records. Personalized Medicine 4(4), 453–455 (2007)
26. Shabo, A.: How can the emerging patient-centric health records lower costs in pharmaco-genomics? Pharmacogenomics 8(5), 507–511 (2007)
27. Implementation Guide for CDA Release 2: Imaging Integration Levels 1, 2, and 3, Basic Imaging Reports in CDA and DICOM Diagnostic Imaging Reports (DIR) – Universal Realm, Based on HL7 CDA Release 2.0, Release 1.0, Informative Document Second Ballot (September 2008)
28. IHE - Integrating the Healthcare Enterprise, http://www.ihe.net/
29. http://wiki.ihe.net/
    index.php?title=Cross_Enterprise_Document_Sharing
30. http://www.haifa.ibm.com/projects/software/ihii/index.html

InEDvance: Advanced IT in Support
of Emergency Department Management

Segev Wasserkrug[1], Ohad Greenshpan[1], Yariv N. Marmor[2], Boaz Carmeli[1],
Pnina Vortman[1], Fuad Basis[3], Dagan Schwartz[3], and Avishai Mandelbaum[2]

[1] IBM Haifa Research Lab
[2] Technion – Israel Institute of Technology
[3] Rambam Health Care Campus

Abstract. Emergency Departments (ED) are highly dynamic environments comprising complex multi-dimensional patient-care processes. In recent decades, there has been increased pressure to improve ED services, while taking into account various aspects such as clinical quality, operational efficiency, and cost performance. Unfortunately, the information systems in today's EDs cannot access the data required to provide a holistic view of the ED in a complete and timely fashion. What does exist is a set of disjoint information systems that provide some of the required data, without any additional structured tools to manage the ED processes. We present a concept for the design of an IT system that provides advanced management functionality to the ED. The system is composed of three major layers: data collection, analytics, and the user interface. The data collection layer integrates the IT systems that already exist in the ED and newly introduced systems such as sensor-based patient tracking. The analytics component combines methods and algorithms that turn the data into valuable knowledge. An advanced user interface serves as a tool to help make intelligent decisions based on that knowledge. We also describe several scenarios that demonstrate the use and impact of such a system on ED management. Such a system can be implemented in gradual stages, enabling incremental and ongoing improvements in managing the ED care processes. The multi-disciplinary vision presented here is based on the authors' extensive experience and their collective records of accomplishment in emergency departments, business optimization, and the development of IT systems.

Keywords: Healthcare, Management, Hospital, ED, Business, Intelligence.

1 Introduction

The rising costs of healthcare services are the result of factors such as increased life expectancy and the availability of an ever increasing number of costly diagnostic and therapeutic modalities. These expenses are imposing pressure on healthcare providers to improve the management of quality and efficiency in their organizations. Both operational and financial aspects must be considered while preserving and improving the clinical quality given to patients [1]. The hospital's Emergency Department (ED) is a

Y.A. Feldman, D. Kraft, and T. Kuflik (Eds.): NGITS 2009, LNCS 5831, pp. 86–95, 2009.
© Springer-Verlag Berlin Heidelberg 2009

prominent scenario in which such managerial improvements are desperately needed, for a number of reasons. First, the ED serves as the main gateway to the hospital and therefore, any inefficiencies and erroneous decisions made will ripple throughout the overall process. Second, the time intervals within which clinical and operational decisions must be taken are short, when compared to those in other departments. Third, due to short horizons in the diagnostic and treatment processes, the time that can be dedicated to data collection is restricted. Fourth, patient flow is highly complex, due to the number and variety of decisions that must be exercised, the diverse skills of caregivers involved in the process, and the wide scope of patients' clinical needs. These special characteristics serve as a basis for the work presented in this paper. They are further elaborated on in the *Current State* section.

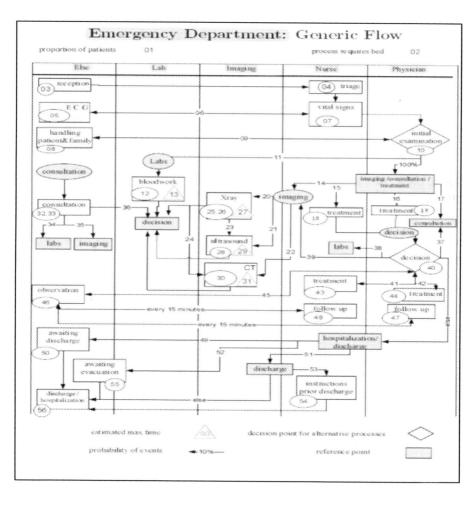

Fig. 1. ED Care Process [1]

In many ways, the requirements for managing ED processes resemble those for process management in other domains. Extensive observations [1,4], both horizontal (snapshots of processes in the ED) and longitudinal (tracking patient flow), exposed three major challenges in the management of ED processes: 1) The ability of the staff (physicians, nurses, managers) to identify the physical location of patients within the process; 2) The ability of staff to obtain patients' current state, both operationally (e.g., waiting for returning tests) and clinically (e.g., severity status); 3) The availability of relevant clinical information in real time. Although there has been significant progress in hospital IT systems (see *Current State* section), the lack of adequate electronic information is still a major obstacle in addressing the above challenges, for hospitals in general and EDs in particular. We view the data collection step as a prerequisite for the development of a process management system for EDs. On top of this layer, an analytics layer should be developed to provide capabilities for identifying bottlenecks, predicting future ED status, and providing recommendations for intervention prior to the occurrence of acute events. On top of these, there should also be a user interface, depicting the relevant information in the best possible manner at the most appropriate time, thus supporting decision-making.

In this paper, we present our vision for an advanced system that can monitor, control, and manage hospital EDs. We describe a system called InEDvance, which implements our vision and enhances existing ED business intelligence systems. (For concreteness, we demonstrate our vision through enhancement of the IBM Cognos BI platform.). We then provide examples showing how this system can be used by ED staff and executives, and discuss its impact on the ED environment. The multi-disciplinary work presented here is based on extensive experience, as manifested through the list of authors and their collective track record in EDs, business optimization, and the development of IT systems.

The rest of the paper starts by describing the unique characteristics of the ED and presenting our vision. Next, we introduce *InEDvance*, a system proposed for implementation in such an environment, and discuss its impact. We conclude with directions for future work and the relevant issues that must be addressed when developing systems such as *InEDvance*.

2   The ED Environment

The environment of the ED is unique and challenging, both clinically and operationally. Patients arriving at the ED are heterogeneous from a clinical viewpoint and suffer from wide variety of medical symptoms. Some patients require urgent treatment while others may wait without any obvious risk to their condition. In short, patients require different customized treatment processes, which increases the operational complexity of the system. For example, the ED must have mechanisms for catering to both ambulatory patients as well as acute patients. Lengths of stay (LOS) at the ED may vary from less than an hour in minor cases to one or two days in extreme cases; therefore, we need to examine processes both horizontally and longitudinally. The ED staff is not familiar with the patients arriving for treatment nor with their medical history; hence, both time and effort must be spent gathering the right information in a timely manner in order to support the decision-making. The traffic inside the ED caused by the movement of staff, patients, and machinery, is heavy and contributes to the complexity of operational processes. It often reaches extreme levels due to the addition of entities that are not part of the process (e.g., companions or visitors of patients); these are difficult to model, monitor, and control. The rate of patient arrival to the ED is also very high, when compared to other hospital departments. Furthermore, arrival rates vary across hours, days, and months.

Due to the high rates of ED arrivals and the difficulty in predicting their numbers (and hence ED workload), overcrowding is a prevalent phenomenon in most EDs [2]. Moreover, the variety of parameters influencing the state of the ED means there is no simple way to predict or even determine when the ED will be, or is, overcrowded. Hoot et al. [3] compared four different methods for measuring ED overcrowding. They reached the conclusion that none of these measures could provide prior warning of overcrowding with acceptable rates of false positives.

As an example for operational complexity, we investigated the environment of the ED at the Rambam Health Care Campus, a major hospital that provides health services to the entire northern region of Israel. Typically, the Rambam ED has over 300 patient arrivals per day, accommodated in 34 beds. Average length of stay (ALOS) is 6.4 ± 2.2 hours. This is relatively short compared to other departments in the hospital, where ALOS of 3.3 ± 1.6 days [4]. An analysis of the arrivals exhibited a diurnal pattern with two peaks at noon and at 7 PM. However, the pattern is influenced by various parameters [4], which impose extreme burdens on the ED staff. This causes deficiencies in data collection and can lead to flawed decision making. Until now, we reviewed the operational and clinical complexity of the ED, detaching it from its IT system. We now add this layer of complexity in the ED environment.

During the last two decades, hospitals in general and EDs in particular have undergone dramatic changes. Many of these changes have been credited to the adoption of Information Technology (IT) systems, often called Hospital Information Systems or Health Information Systems (both referred to as HIS) [5,6,7]. The HIS includes components that are responsible for managing the clinical status of patients along the care process. These systems, usually referred to as Electronic Health Records (EHR), are disconnected from other systems inside the ED and outside the hospital. Unfortunately, they lack basic capabilities, which causes them to fall short even in the stages of development and deployment; [8] inspected collaboration aspects which were found necessary for such environments.

Most HIS development efforts are invested in clinical IT systems, while fewer efforts are devoted to developing solutions for monitoring, management, and control. For such controls to be effective, we need various metrics in the ED. Better measurement of basic operational indicators, such as the individual patient's LOS, is crucial for improving ED care processes. There are three main challenges in providing such indicators. First is the ability to measure them in the ED. Second is the ability to define which indicators can contribute to process efficiency. Third is the ability to provide these measurements in a way that is clear and useful for ED staff and executives, enabling them to react appropriately.

Regarding the first challenge, we believe it is critical to know the physical location of patients, physicians, and other healthcare personnel, at every point in time. New improvements in RFID technology are paving the way for accurate location tracking, which can provide valuable insight into the operational processes within the ED and the patient's status within these processes. Still, these systems provide only raw data, which must still be processed by other event-based systems (i.e., CEP, see *Vision* section) or be integrated with traditional systems inside the hospital, such as EHRs [14], lab results management systems, and imaging systems.

Regarding the second challenge, we need to process the raw data and make it available to measure Key Performance Indicators (KPIs). This has the potential to improve the process of patient flow, notably waiting times, treatment times, and queue lengths. After these KPIs are defined, we must identify the optimal way to display the information, for each and every consumer and scenario.

Although it is difficult to measure the impact of IT systems on the ED environment, reviews on health information systems have shown that treatment quality improves with increased adherence to guidelines, enhanced disease surveillance, decreased medication errors [9], and increased patient safety [10]. This has been a field is of great research interest [11,12].

The next section describes an advanced IT system that we designed and believe can cope with the above-mentioned difficulties.

3 Vision

The concept of hospital Emergency Departments (EDs) has evolved over the centuries. Their origins stem from the experience of military medical care during war times and therefore lacked advanced processes to care for the aged and chronically ill. From simple caregivers, modern hospitals have evolved into becoming leaders in the development and delivery of care to the community. As the leading environment for health professional education, they are actively engaged in medical research and the promotion of good health [13].

As already noted, the ED serves as the main gate for patients and is their first encounter with hospital care services. Thus, the need for improvement across all dimensions (clinical, operational, and financial) is crucial and must be continuous.

3.1 System Description

The decision support system for the management of a hospital ED should contain the following main components (See Fig 2):

Data Collection Component: The Data Collection component is responsible for gathering data of various types and consists of a set of information systems. This includes all data currently being collected in the ED (e.g., EMR [14], imaging, ERP). In addition, it should incorporate additional information, such as that coming from RFID tracking systems [15] for patients and instruments [16][17], and record the interaction times between patients and medical staff. Such a system must be able to collect relevant information without imposing the additional burden on the staff. RFID tracking has the ability to increase efficiency, and reduce costs and medical errors due to erroneous identification of patients. Although such tracking systems are very popular in other domains, hospitals in general and EDs in particular have not yet adopted this type of system. We are exploring the field, but further details are beyond the scope of this paper. In addition to the data being collected inside the ED, the Data Collection

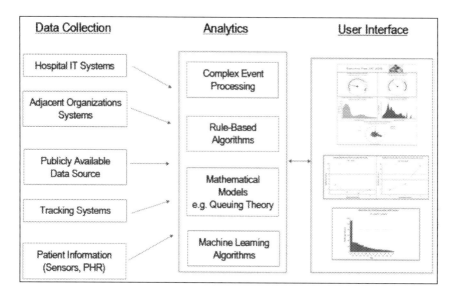

Fig. 2. InEDvance System Architecture

component will gather information being imported from other healthcare providers (e.g., other hospitals and clinics), from adjacent organizations (e.g., insurance companies) and from patients (e.g., sensors collecting data). Moreover, the component will gather publicly available information from the web, such as one providing alerts on adverse drug reactions, publications, and so forth.

Analytics Component: The Analytics Component enables advanced analysis methods, including the following:

- Data consolidation, cleansing and correlation - Hospital data collected and produced by IT systems is often noisy, and needs to be consolidated and correlated based on additional knowledge and/or assumptions. Take, for example, raw data produced by a tracking system that provides information on the location of patients at each point in time. Now consider two consecutive location events, one indicating that the patient has entered a treatment room and another indicating that the patient has left the room. If the time difference between the two events is ten seconds, obviously, the patient did not receive any treatment in the room. However, if the time difference between the two events is thirty minutes, one could like to conclude that treatment was provided to the patient. Techniques such as *Complex Event Processing (CEP)* [18] can be used to integrate these conclusions.
- Forecasting algorithms – These can be used to predict future events (e.g., patient arrivals) or inferred knowledge (e.g., load on the ED staff) to support smarter decision making.

- Mathematical models – For example, queuing equations or simulation models can be used to model the operational aspects. *Markov Decision Process* (*MDP*) models can be used to model clinical decision making [19].
- Data completion algorithms - These will provide data that cannot be delivered by the existing IT systems, to ensure a complete picture of the state at the ED.
- Optimization algorithms - These will enable optimal decision making in various and intersecting dimensions, i.e., clinical, operational and financial.

User Interface Component: In order to overcome the challenges involved in deploying IT systems in hospitals [8], the user interface must suit the nature of the ED and support online decision making, with high flexibility for events and situations. It will also need to have several different methods for presentation (e.g., online/offline) with varying resolution and depth of data consumption. The user interface component will enable the following:

- Dashboard for the minimal subset of information to support quick decision making. This will be utilized in an "online" mode by the ED staff
- Comprehensive reporting capabilities enabling complex queries. This will include a combination of graphs and data.
- Advanced data mining capabilities to turn the data into decisions, such as identification of dominant factors causing an observed state at the ED.
- Adaptive mechanism to dynamically change the data and applications being consumed (like done in Web Mashups [20]), change the way they are connected in real time [21] using highly efficient algorithms, and adjust them according to the changing context and situation of the user by adopting methods from artificial intelligence and planning [22].

We consider a design following the principles of the Service Oriented Architecture approach [23] and open standards to ensure that the infrastructure is flexible and modular.

In order for this vision to be achieved, basic research is required in several areas (e.g., sensor networks, data correlation) and is also under our ongoing exploration as well. In addition, an adjustment of formal models and methods should be done to fit the ED environment, with its special characteristics.

3.2 Scenarios

We now describe several real-life end-to-end scenarios showing how such a system can contribute to the management of a hospital ED. We differentiate between two types of scenarios that are of great daily interest. Each scenario is investigated from three perspectives: i) Definition of user needs ii) Selection of metrics and Key Performance Indicators (KPIs), iii) Interface design. The first focuses on the features the user would like to benefit from, and on its role. The second focuses on the set of KPIs that are about to be examined and on the way to measure them. The third focuses on the identification of the most usable display to be shown to the user (e.g., the level of details) and on the interaction (e.g., drill down capabilities) required by the user in such a scenario. The first scenario demonstrates an offline analysis done by an ED executive at large time intervals (e.g., every month). The second scenario is done online by physicians in a shift. The examples provided are demonstrated by the InEDvance system we implemented at the IBM Haifa Research Lab, enhancing the infrastructure of IBM ® Cognos platform. Screenshots provided in Fig 3 are taken from this described system.

Offline Analysis: The first scenario discusses a process that can be done by ED executives on a monthly basis or when required. (i) Definition of user needs: Since the rate of patient arrival at the ED changes over time, it is difficult to determine the optimal amount of beds for any point in time. Managers can more easily determine the distribution of bed occupancy at each time of day and day of week (DOW) if they know the rate and pattern of arrivals, the flow of each patient along the process, and the steps in the process that require a bed. Having a larger number of beds will result in lower bed utilization and will force the staff to keep moving patients to avoid long walking distances between consecutive treatments. A lower number will result in longer waiting times for available beds. (ii) Metrics and KPIs: We calculate the full distribution including the average, 90th-percentile, and 95th-percentile of beds occupied at each hour and on each day of the week. A recommendation given by the system, based on the analyzed data, will help executives understand the consequences of each decision. (iii) User Interface: We show the user the system recommendation for 90th-percentile for each hour and DOW with a maximum number of beds being required for each. We also allow the user to enter a fixed number of beds, and ask the system to calculate the risk of having a shortage in amount of beds for each hour and DOW. Fig 3(a) shows an example of such a view, with alternative displays. The gauge charts present the averaged value over a month, while the line charts present the distribution over the hours in a day.

Online Analysis: The second scenario demonstrates an online consumption of data by physicians in a shift. (i) Definition of user needs: In most large hospitals, there are several imaging rooms, one of which is dedicated to the ED. The physician often decides whether to direct a patient to the abovementioned dedicated room or to a remote one. If the patient is sent to an overloaded room, it will result in a long waiting time. For intelligent decision making, the physician needs to know the future load of each location. Most ED physicians use rules of thumb that are hard to follow and model. A more accurate prediction of future load for each room would help shorten patient waiting times in the imagining room and improve the quality of care. (ii) Metrics and KPIs: There are two different KPIs needed here. The first is the predicted waiting time for the patients who arrive at each imaging room. The second is the impact of the decision to direct each patient to a specific room. For that, we need to predict how many patients will arrive at each room before our current patient gets there, and to estimate the time required for their treatment. An average and 90th-percentile is sufficient in a real-life scenario. We also need to understand which processes the patient is involved in and their current state. For example, if the patient's lab results are only expected to be ready after the patient returns, it would be reasonable to send the patient to the loaded room, unless prohibited due to clinical risks . (iii) User Interface: We use side-by-side approach, showing the prediction of waiting times for the next patient for each room, next to a list of processes with their states for the current patient. For each process, we show its starting time, followed by its expected time (average and the 90th-percentile). Fig 3(b) shows an example of an online view integrating the current state in the ED shift and a prediction of the expected state for the subsequent hours.

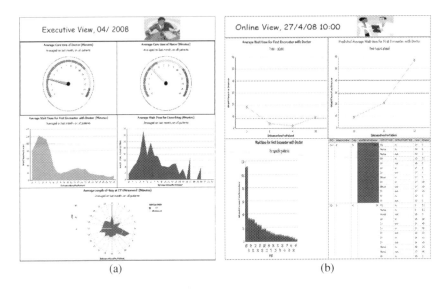

Fig. 3. InEdvance Reports for (a) executives and (b) shift personnel

4 Summary and Future Work

In this paper, we reviewed the complexity of the ED environment from various aspects, and elaborated on the IT obstacles associated with trying to equip the ED staff with systems that can improve management capabilities and performance measurement. We also presented our vision of a system that can optimize care, followed by examples demonstrating the improvement of service quality given at the ED, reduction in pressure on the ED staff and reduced cost. The vision presented in this paper paves the way to various research directions that involve IT and adjacent domains. Due to limited space, we will give several example research questions that we find of great interest: (1) How can we perform intelligent integration of real-life data of various types with simulation and forecasting algorithms. (2) How can we deal with noisy data, such as that coming from RFID systems. (3) What is an appropriate set of KPIs and design for each type of system client (physician, nurse, patient).

References

1. Sinreich, D., Marmor, Y.: Emergency Department Operations: The Basis for Developing a Simulation Tool. IIE Transactions 37(3), 233–245 (2005)
2. Gibon, C.: Reducing Hospital Emergency Department Crowding, White Paper
3. Hoot, N.R., Zhou, C., Jones, I., Aronsky, D.: Measuring and forecasting emergency department crowding in real time. Ann. Emerg. Med. 49(6), 747–755 (2007)
4. Basis, F., Pollack, S., Utits, L., Michaelson, M.: Improving the pattern of work towards emergency medicine at the Emergency Department in Rambam Medical Centre. Eur. J. Emerg. Med. 12(2), 57–62 (2005)

5. Prokosch, H., Dudeck, J.: Hospital Information Systems: Design and Development Characteristics; Impact and Future Architecture. Elsevier, Amsterdam (1995)
6. Ball, M.J.: Hospital information systems: perspectives on problems and prospects, 1979 and 2002. Int. J. Med. Inform. 69(2-3), 83–89 (2003)
7. Kuhn, K.A., Giuse, D.A.: From Hospital Information Systems to Health Information Systems Problems, Challenges, Perspectives. Methods of Information in Medicine 40(4), 275–287 (2001)
8. Pratt, W., Reddy, M.C., McDonald, D.W., Tarczy-Hornoch, P., Gennari, J.H.: Incorporating ideas from computer-supported cooperative work. J. of Biomedical Informatics 37(2), 128–137 (2004)
9. Chaudhry, B., Wang, J., Wu, S., Maglione, M., Mojica, W., Roth, E., Morton, S.C., Shekelle, P.G.: Systematic review: impact of health information technology on quality, efficiency, and costs of medical care. Ann. Intern. Med. 144(10), 742–752 (2006)
10. Menachemi, N., Saunders, C., Chukmaitov, A., Matthews, M.C., Brooks, R.G.: Hospital adoption of information technologies and improved patient safety: a study of 98 hospitals in Florida. J. Healthc. Manag. 52(6), 398–409 (2007); discussion 410
11. Devaraj, S., Kohli, R.: Information technology payoff in the health-care industry: a longitudinal study. J. Manage. Inf. Syst. 16(4), 41–67 (2000)
12. Heathfield, H., Pitty, D., Hanka, R.: Evaluating information technology in health care: barriers and challenges. BMJ 316(7149), 1959–1961 (1998)
13. Reiling, J.: Safe by Design: Designing Safety in Health Care Facilities, Processes, and Culture; 1008 isbn, 978-1-59940-104-1; Published by Joint Commission Resources, and available through, http://www.jcrinc.com/
14. Mon, D.T.: Defining the differences between the CPR, EMR, and EHR. J. AHIMA 75(9), 74–75, 77 (2004)
15. Das, R.: An Introduction to RFID and Tagging Technologies (Cambridge, UK: White paper, IDTechEx, 2002) (2002), http://www.idtechex.com (accessed November 23, 2005)
16. Shih-Wei, L., Shao-You, C., Jane, Y.H., Polly, H., Chuang-Wen, Y.: Emergency Care Management with Location-Aware Services. In: Pervasive Health Conference and Workshops 2006, pp. 1–6 (2006)
17. Miller, M.J., Ferrin, D.M., Flynn, T., Ashby, M., White, K.P., Mauer, M.G.: Using RFID technologies to capture simulation data in a hospital emergency department. In: Perrone, L.F., Lawson, B.G., Liu, J., Wieland, F.P. (eds.) Proceedings of the 38th Conference on Winter Simulation, Monterey, California, December 03 - 06, 2006, pp. 1365–1370 (2006)
18. Etzion, O.: Complex Event Processing, Web Services. In: IEEE International Conference on Web Services, ICWS 2004 (2004)
19. Brafman, R.I., Domshlak, C., Shimony, S.E.: Qualitative decision making in adaptive presentation of structured information. ACM Trans. Inf. Syst. 22(4), 503–539 (2004)
20. Simmen, D.E., Altinel, M., Markl, V., Padmanabhan, S., Singh, A.: Damia: data mashups for intranet applications. In: SIGMOD 2008 (2008)
21. Abiteboul, S., Greenshpan, O., Milo, T., Polyzotis, N.: MatchUp: Autocompletion for Mashups. In: ICDE 2009 (2009)
22. Brafman, R.I., Domshlak, C., Shimony, E.: Qualitative decision making in adaptive presentation of structured information. ACM Trans. Inf. Syst. 22(4), 503–539 (2004)
23. Chu, S.C.: From component-based to service oriented software architecture for healthcare. Journal on Information Technology in Healthcare 4, 5–14 (2006)

Enhancing Text Readability in Damaged Documents

Gideon Frieder

Documents can be damaged for various reasons – attempts to destroy the document, aging, natural cause such as floods, etc. In the preliminary work reported herein, we present some results of processes that enhance the visibility of lines, therefore the readability of text in such documents. No attempt is made to interpret the contents – rather, the work intends to aid an analyst that will eventually process the information that is now easier to see and acquire.

Our software provides different methods of enhancement, and the analyst can choose the method and set the optimal parameters. The results that we have are rather encouraging, however, the processing is done on every page separately, and the choice of parameters to achieve optimal enhancement is difficult. The major obstacle for automating this process and providing a tool which can be used for a reasonable – even massive – amount of data lies in the absence of a mathematical definition of readability, and ability to create a figure of merit whose value will guide the automatic selection of optimal parameters. We shall present our results, and hope that it will motivate the listeners to help in finding such figure of merit.

The corpus of data used in this presentation consists of the diaries of HaRav Dr. Avraham Abba Frieder. These diaries consist of more then 800 pages written in different scripts, both typed and handwritten, in different languages, in different states of readability; with intermix of text, pictures, tables, corrections etc.

The diaries are now on a permanent loan to the archives of Yad Va Shem, and can be found on the net in www.ir.iit.edu/collections

This work was initiated while the presenter was visiting the imaging laboratory of the University of Bremen. It was continued and greatly expanded by Dr. Gady Agam and his collaborators in the IR@IIT Laboratory, under the direction of Dr. Ophir Frieder.

ITRA under Partitions

Aviv Dagan[1] and Eliezer Dekel[2]

[1] Haifa University
avivdagan@gmail.com
[2] IBM Haifa Research Laboratory
dekel@il.ibm.com

Abstract. In Service Oriented Architecture (SOA), web services may span several sites or logical tiers, each responsible for some part of the service. Most services need to be highly reliable and should allow no data corruption. A known problem in distributed systems that may lead to data corruption or inconsistency is the partition problem, also known as the split-brain phenomena. A split-brain occurs when a network, hardware, or software malfunction breaks a cluster of computer into several separate sub-clusters that reside side by side and are not aware of each other. When, during a session, two or more of these sub-clusters serve the same client, the data may become inconsistent or corrupted.

ITRA – Inter Tier Relationship Architecture [1] enables web services to transparently recover from multiple failures in a multi-tier environment and to achieve continuous availability. However, the ITRA protocol does not handle partitions. In this paper we propose an extension to ITRA that supports continuous availability under partitions. Our unique approach, discussed in this paper, deals with partitions in multi-tier environments using the collaboration of neighboring tiers.

Keywords: partitions, split-brain, multi-tier environment, SOA, web services, ITRA.

1 Introduction

Nowadays, as the internet is available almost everywhere, we are witness to its rapid evolvement in several fields: for personal, business, and scientific use. More and more web services are being offered to internet users; for example, book purchases at Amazon, auctions at eBay, finance services, supermarket shopping, and social networks.

The simplest service offered is built from two end points: the *client*, who consumes the service, and the *server*, who supplies it. A common form of communication between the client and the server uses service representatives, also called *stubs*. When the client needs a service, it "requests" a stub from the server. The server is responsible for allocating the required resources, creating the stub, and sending it to the client. As the client gets the stub it communicates with the service using that stub only. This mechanism is called a *proxy* and it is very popular in distributed systems. The proxy simplifies the client's work since the client activates the proxy representative as if it

Y.A. Feldman, D. Kraft, and T. Kuflik (Eds.): NGITS 2009, LNCS 5831, pp. 97–108, 2009.
© Springer-Verlag Berlin Heidelberg 2009

was the service itself. It also maintains the object oriented encapsulation principle since the stub does not reveal the service's inner structure; only the interface is accessible. Examples of the proxy design pattern can be found in Java-based distributed systems such as Java RMI and JINI and in CORBA-based systems.

When referring to a serving system, it probably consists of a cluster (or clusters) of computers that together form the system. The cluster members, also called *nodes*, are connected to one another and in most cases are all equal and have the same role; to serve clients. Having a cluster of computers has several advantages. A cluster of computers can simultaneously serve many clients; the more nodes a cluster has, the more clients (end users or other nodes) it can simultaneously serve. Also, since computers may fail every once in a while, working with nodes can assist in that a working node can take over a failed one without the client being aware of it.

Observing today's web services shows that in many cases the service lifecycle includes transferring data between several logical levels, called *tiers*. Each tier is, in most cases, a cluster of computers that serves another tier.

An interesting and problematic phenomenon that may occur in a multi-tier environment is *partitioning*, also known as the split-brain phenomenon. A partition is created when, due to a network malfunction, a cluster is split into two or more groups of nodes. Each group cannot communicate with the others but can communicate with the world outside the cluster. In other words, each group acts as if only its members form the cluster. A situation where multiple partitions exist side by side can lead to data corruption that may cause a disaster, for example multiple credit card charges in a business web service.

ITRA – Inter Tier Relationship Architecture [1] enables web services to transparently recover from multiple failures in multi-tier environment and to achieve continuous availability. However, the ITRA protocol does not address the issue of partitions. In this paper we propose an extension to ITRA that supports continuous availability under partitions. Our unique approach deals with partitions in a multi-tier environment using collaboration of neighboring tiers.

The following chapter discusses partitions and shows some of the existing different solutions. Chapter 3 introduces ITRA. Chapter 4 shows the possible scenarios under ITRA and Chapter 5 deals with the software simulation implementation and its results.

2 Partitions

The partitioned or split-brain phenomenon is a well known problem in the field of distributed systems. A partition is formed as a result of a network failure, causing a disconnection between groups of cluster members. Since each partition is not aware of the problem, it acts as if its members are the only ones existing and that they are the only ones who form the cluster. All partitions have a connection to the world outside the cluster so a situation where two or more partitions residing side by side and serving clients is feasible. This situation can cause a real damage in the form of data corruption. An example may be shopping in a partitioned online store. A user is browsing a store with two partitions A and B. He chooses a product and goes to pay for it. The user fills the necessary forms and presses the "purchase" button. The order action causes the system to allocate a node to handle the order. The serving node is chosen from partition A, which performs the order and sends the response (a confirmation web page) to the user. Unfortunately at this point, before the page is presented on the user's browser, the user's browser fails and restarted. After a short while, when the browser is back online, the user re-enters the store. Since the user did not receive the confirmation, he enters the "My Purchases" section to see if the item he chose is there. Unhappily, the serving node is now from partition B, which is unaware of the recent purchase. Thus, the user does not see the item in the list and performs the process again, causing duplicate credit card charges.

Several solutions can be found in the market in off-the-shelf products, and they are reviewed later in this chapter. Some solutions try to lower the chances of partitions forming, while others try to overcome the problem. The ultimate solution should include both approaches. It should be mentioned here that all the solutions discussed are at the level of the cluster. None of the approaches handles partitions using the collaboration of neighbor tiers, as we suggest in this paper.

2.1 Multiple Links

The multiple link solution is based on the assumption that two nodes of a cluster might disconnect, but with a low probability. If we take a cluster with only two nodes and connect them with several independent links, the disconnection probability drops, since the nodes disconnect only if all links fail.

The links between the nodes can be regular LAN connections, but less conventional ways are also possible; for example, using a shared disk. In general, most clusters have at least two separate networks, as all nodes have two network cards (NICs), each connected to a different network. As we have said, this method lowers the chance of partitions occurring but it cannot totally avoid them. Also, maintaining this kind of system may be expensive for large systems.

2.2 Master Node

In some cluster-based systems, a specific node is always a member of the cluster. This node can act as a master node in that its job is to build and manage the cluster. All nodes know who the master is. When nodes want to join the cluster, they address a request to the master node. Since the master itself is not failure proof, it probably has a backup node (or a few backups) or else a master node failure disables the entire cluster. In this way, partitions do not occur; any node that cannot connect the master node is not a member of the cluster. However, this solution is limited. First, there is always a chance that the master node and its backups will all fail, which fails the whole cluster. Second, it is possible that no nodes can connect to the master (but can connect to each other) so only the master node forms the cluster.

2.3 Lock

The lock solution is based on a global resource, acting as a lock, which is available to all cluster members; for example, a shared disk. Any node may check the lock (to see if it is locked or unlocked), but only one can claim ownership and prevent the others from taking ownership. The locking system allows dynamic cluster management. In every membership update of the cluster, the nodes try to take ownership of the lock. When a node wants to join the cluster, it checks the lock status. If it is unlocked, the node locks it and becomes the cluster's master node. If the lock is already locked, the node checks who is the lock owner and requests to join the cluster. To avoid a situation where the lock owner is terminated and therefore no other node can become the master, the locking system includes a watchdog process, as follows: as the node claims ownership, a countdown timer is set. The master node must reset the timer before the time is up, or else the lock releases itself, thereby allowing another node to become the cluster master node. A split brain does not happen when using this solution. If the cluster is split, all nodes try to get the lock. If the master is still alive, only his sub-cluster forms the cluster; otherwise, any one of the sub-clusters can form the cluster. Even though the odds are small, a situation of a singleton that succeeds in taking the lock and forming the cluster is possible. Another weakness is the dependency on the locking service itself; if, for some reason, the service fails, the whole cluster becomes unavailable.

2.4 Quorum Service

The quorum service solution solves the problem introduced above, where a one-node (or few-node minority) cluster is created. The idea behind this solution is that the cluster is built from a majority node set. The cluster configuration, or the membership map, is stored in a shared resource or replicated over the cluster nodes. In cases where the cluster is built of many nodes or spread over a large geographic area, storing the map on each node's storage is more efficient.

Initially, when the first node starts, it tries to join the cluster but fails since the cluster is not yet formed. The node "hibernates" for a while and tries again later. Only after at least half of the nodes are up and can communicate with each other, the cluster is configured and the membership map is replicated on each node. If, at some point, the cluster loses a majority of nodes, the cluster becomes unavailable.

On a partitioning event, only a partition that has a majority continues to work as a cluster. The others partitions will not form a cluster. However, what happens if the cluster is partitioned to a set of partitions, each with less than a majority node set? Or what if two partitions have exactly half of the nodes each? We suggest some enhancements. A locking system can be used as a tie breaker in the case of two half-sized partitions. A weighted majority means that each node is weighted according to its failure rates (based on statistics) so that more reliable nodes are assigned a higher weight. And when the cluster splits into several partitions, the partition with the highest weight forms the cluster.

2.5 Quorum Server

The quorum server is similar to the quorum service, but its implementation is outside the cluster. When the cluster is formed, all nodes access the server and only when a majority is reached is the cluster created. The enhancements described in the previous section can be applied to the quorum server too.

2.6 Arbitration Nodes

Clusters may be built in a multi-site configuration, called *metro-clusters*. One reason to use this architecture is to achieve a highly available system even in the presence of site disasters. Each site contains several nodes that together form the cluster. A network disconnection between sites can cause a split-brain (for example, when the cluster is built from two sites). The Arbitration nodes solution can be used in this kind of configuration. Arbitration nodes are a small number of computers, located in another site, that together with the rest of the nodes form the cluster. As members of the cluster, they can communicate with all other nodes. The use of arbitration nodes reduces the chance of a split-brain occurrence. As long as there is a path between sites—such that the path can pass through the arbitration nodes—the cluster does not partition. Although it is possible for the arbitration nodes to act as regular nodes of the cluster, this behavior is not recommended as it can cause a bottleneck. In most cases, the arbitration nodes act as an extension of the quorum server or a server locking system. When needed, the arbitration node decides which site becomes a member of the cluster. Weaknesses of the arbitration node architecture are that it provides no solution for a partition inside the site, and that the arbitration nodes may fail, thereby causing a partitioned system.

2.7 Paxos

Decisions in distributed systems may be taken using the Paxos consensus protocol. The main idea behind Paxos is that decisions are made by a leader who is chosen by a majority of participants. Paxos defines three roles: Proposers are nodes that are proposing decisions, Acceptors receive the proposals, and Learners get the final decision and act accordingly. Any decision may be taken only if a majority of Acceptors agree on the one proposal. Any node can serve in any of the roles. Roles might change as the protocol progresses.

Making a decision in Paxos is a two-phase process and each is divided into two subphases.

- 1A. Prepare – The proposers chooses a serial number N and send a Prepare message to the acceptors with that proposal number (the proposers choose their own serial number).
- 1B. Promise – The Acceptors receive the Prepare messages. When an Acceptor receives a Prepare message with N larger than all former serial numbers, it sends a Promise message to the Proposer that it cannot accept any proposal with a serial number less than N. If the acceptor has already accepted a

proposal, it attaches the value it accepted. The value is actually the decision; for example, who will serve an incoming request.

- 2A. Accept – When a Proposer receives a Promise message from a majority of Acceptors, the Proposer may choose a value from the set of accepted values. If the set is empty, the Proposer can decide to choose any value and sends the Acceptors an Accept message with that value.

- 2B. Accepted – An Acceptor that receives an Accept message accepts the attached value unless the Acceptor has already promised a higher N to another Proposer. If the Acceptor accepts the proposal, it sends an Accepted message to the Proposer and all learners.

There are several versions of the Paxos protocol; All the Paxos protocol implementations can withstand partitions. The weakness of Paxos is the possibility of a multi-group partition where none of the partitions has a majority quorum.

3   ITRA – Inter Tier Relationship Architecture

The ITRA - Inter-Tier Relationship Architecture was first introduced in [1] and its main characteristics are described in this section. The main concept of ITRA is inter-tier collaboration to achieve high availability multi-tier systems that can detect failures and overcome them transparently, without human aid.

Figure 1 describes how two tiers communicate with each other. It includes some notations that we use in this paper. As shown, the communication is via stubs. When a node x from tier k-1, denoted as $n_{x_{k-1}}$, wants to activate the next tier t_k, it first needs to receive a t_k stub. t_{k-1} sends the set of operations against t_k denoted as O_k, in sequential order, meaning that in case of two operations O_{k_i} that is former to O_{k_j}, the latter is sent only after O_{k_i} is complete. The figure indicates that $n_{x_{k-1}}$ does not know t_k which node serves it, as this is part of the tier encapsulation property.

The ITRA mechanism comprises an operation tagging system. This tagging system assigns each operation a unique sequential ID. These IDs are reproducible. Thus, a given operation is assigned with the same sequence number, regardless of the particular node that executes it. The ITRA tagging system allows easy tracking of operations; for example, searching the logs of each tier may be based on the operation ID as the searching key. The sequence number contains numbers delimited by dots. Each part of the sequence number represents a tier in the computation chain.

Following are some more notations for our discussion:

- s_k – A sequence number of an operation against t_k. s_{k_i} - The sequence number of operation i against t_k.

- $s_{k_i} < s_{k_j}$ – The sequence number of operation i is lexicographically smaller than the sequence number of operation j. The operators '=' and '>' act in the same manner. If $s_{k_i} < s_{k_j}$ stands, then $O_{k_i} < O_{k_j}$ also stands.

- ss_{k_i} – The suffix of O_{k_i} that was added by the t_k stub.

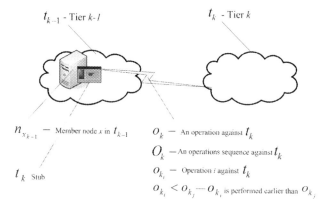

Fig. 1. Inter-tier communication

When operation O_{k-1} produces a set of operations O_k, these operations have the same prefix s_{k-1} and a unique suffix for each operation. The suffix is concatenated to the prefix when a dot is placed in between them. For example, the sequence number for O_{k_i} is $s_{k-1} \cdot ss_{k_i}$. According to the tagging system, the stub keeps the last prefix it received (marked with Lpref and the last suffix that was created, marked with Lsuf). When a stub begins its lifecycle, Lsuf is set to zero. If t_k stub gets an n operation tagged with s_{k-1} that is lower than Lpref, a failure result is returned. The prefix should stay the same or grow by 1. However, if s_{k-1} is equal to Lpref, Lsuf advances by 1. When s_{k-1} is larger than Lpref, Lsuf is set to zero and Lpref is set to s_{k-1}.

4 Possible Scenarios under ITRA

This chapter includes an overview of possible scenarios that may happen in a multi-tier environment system using ITRA architecture and examines whether problems (such as data corruption) caused by partitioning occur in each one. The figures in this section are sequence diagrams showing tiers relevant to the scenarios and events. In most cases, t_k is positioned in the middle surrounded by its neighbors: t_{k-1} from the left and t_{k+1} from the right. The first two scenarios are covered in [1].

4.1 Normal Operation

The normal operation scenario is trivial. The system has no partitions and no connection failures. It can be seen that somewhere along the timeline, as part of, O_k t_{k-1} sends O_{k_i} to t_k. Then, t_{k-1} waits for a result (indicated by the gray broken line) before it sends the next operation in the sequence, O_{k_j}. Tier t_{k-1} is not aware that O_{k_i} involves also t_{k+1}. Tier t_k sends n+1 operations to t_{k+1} (again, each operation is sent only after a result for the prior operation is received). After handling all n+1 operations, t_k returns the result to tier t_{k-1}, which continues the task and sends O_{k_j}.

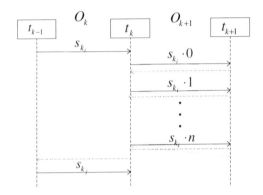

Fig. 2. Normal operation

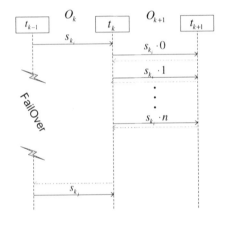

Fig. 3. Failover

4.2 FailOver

The failover scenario starts similarly to the normal operation scenario, by sending t_{k-1} to t_k and sending the operations from t_k to t_{k+1}. Somewhere in the timeline, O_{k_i} suffers from a network failure that causes the active node $n_{x_{k-1}}$ to halt. As seen in Figure 3, this failure has no effect on t_k and t_{k+1}, which continue working. After a while, t_{k-1} overcomes the failure and allocates a new working node $n_{y_{k-1}}$, which takes the place of $n_{x_{k-1}}$. Overcoming this kind of failure is called *Failover*. When $n_{y_{k-1}}$ is connected to t_k, it may get an updated state, indicating that it should wait for a result for O_{k_i}. Situations in which $n_{y_{k-1}}$ sends O_{k_i} again, for example, when the received state is not up to date, are valid due to the non-idempotent ITRA property. Tier t_k notices that it is already serving O_{k_i} and does not start handling it again. If the failover process takes longer than the time it takes t_k to handle O_{k_i}, when $n_{y_{k-1}}$ sends O_{k_i}, Tier t_k immediately returns the pre-calculated results.

4.3 Partition and Failure in t_k

The partition and failure in t_k case starts when t_k is already partitioned. n_{A_k} and n_{B_k} are two nodes from two partitions. At the beginning of the timeline, both have the same states. Tier t_{k-1} starts by sending a part of the operations sequence O_k, which is handled by n_{A_k}. Somewhere in the timeline during the handling of O_k, due to a failure in t_k, n_{B_k} becomes the serving node. When t_{k-1} reconnects to t_k, the t_k stub notices that the state it holds is more updated than the t_k state (since n_{B_k} has not received any of the n_{A_k} updates). The t_k stub feeds tier t_k with its state and continues to send the rest of O_k. If this state feeding process has not been implemented, n_{B_k} expects to receive O_{k_i} but instead gets an operation "from the future" $O_{k_{j+1}}$. If that is the case, n_{B_k} must declare that t_k is partitioned.

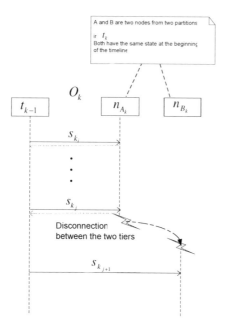

Fig. 4. Partition and failure in tier k

4.4 Re-invocation of Client Synched Operation

The timeline of the client synched scenario begins with a partitioned tier t_{k-1}. Nodes $n_{A_{k-1}}$ and $n_{B_{k-1}}$ are two nodes from different partitions: A and B. $n_{A_{k-1}}$ starts by sending O_{k_i} and eventually receives the answer from tk. Let us say that O_{k_i} is the only operation against t_k. After getting the response for O_{k_i}, $n_{A_{k-1}}$ synchs the t_{k-1} nodes; however, since it is partitioned, only the nodes in partition A receive the update. Node $n_{A_{k-1}}$ then informs t_k that O_{k_i} is client synched. At this point a failover occurs in t_{k-1}, after which $n_{B_{k-1}}$ becomes the serving node replacing $n_{A_{k-1}}$. Since $n_{B_{k-1}}$ is not up to date and it has no result for O_{k_i}, it sends it again to t_k. Without our extension, ITRA handles O_{k_i} as if it has never been served before, since t_k has already erased the history log for O_{k_i}. This might lead to data corruption. However, in our solution, tier t_k notices that it has just received an operation that has been declared as client synched and would inform t_{k-1} that it has partitions. This scenario is also equivalent to situations where t_{k-1} splits just before node $n_{A_{k-1}}$ performs the synch.

4.5 "Swing"

Similar to the client synched scenario, the swing scenario also starts with a partitioned tier; this time, t_k. The nodes n_{A_k} and n_{B_k} are two nodes from different partitions: A and B. t_{k-1} sends O_{k_i} to its serving node, which is n_{B_k}. While handling O_{k_i}, n_{B_k} sends the operation set O_{k+1}. In the meantime, as a result of a failure, t_{k-1} performs a failover that makes n_{A_k} the serving node of t_{k-1}. Since t_{k+1} has not received the result for O_{k_i}, it is sent again and in turn n_{A_k} resends O_{k+1}. The outcome is that t_{k+1} serves two instances of t_k. On the one hand, t_{k-1} receives new operations that need to be handled, and on the other hand it receives "old" operations from n_{A_k}

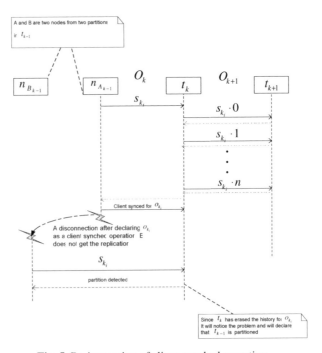

Fig. 5. Re-invocation of client-synched operation

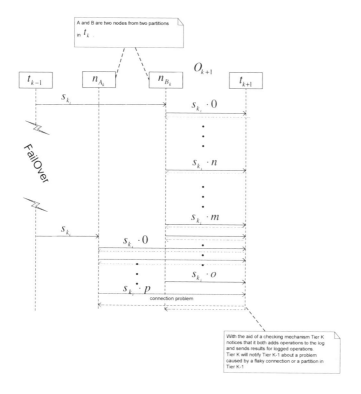

Fig. 6. Swing

whose results can be found in the log. By adding a checking mechanism to ITRA and defining a policy, this situation—where some operations are added to the log and others are extracted from it simultaneously —may be detected and treated by t_{k+1} as a network problem. When t_{k+1} detects the problem, it should notify t_k , or else a re-invocation of the clients synched operation may take place. For example, when n_{A_k} is done handling O_{k+1} (before n_{B_k}) it may send a client synched message to t_{k+1}, which cleans its log, causing n_{B_k} to send a client synched operation.

5 Software Simulation Implementation + Conclusions

Observing the possible scenarios of ITRA under partitions showed that critical scenarios that are sensitive to partitions and may lead to data corruption are those which involve re-invocation of client synched operations. Therefore, a good solution detects that this event is about to happen and takes further actions such as notifying an operator on a control panel that a partition is detected or sending a "suicide" message to the detected partition (terminating a serving partition does no harm since ITRA stubs detect the termination and address their requests to active nodes). To test our approach, we built ITRA simulation software that includes the partitions detection solution. The software simulates a full environment, meaning that several clusters communicate with each other according to the ITRA protocol; while, in the background, network failures and partitions may occur.

The simulation is written in Java using DESMO-J [3], where entities such as clusters, nodes, and stubs are represented by threads. When the simulation runs, it generates the set of operations for the session, executes them, and fails nodes and creates partitions on a random basis. All activities during the session are logged to an HTML output file for further analysis. The software has a set of parameters that control the behavior of the software. For example, the user may change the number of clusters in the systems, change the frequencies of failures and partitions, or choose the pseudo random seed for the entire simulation. Full documentation about the software can be found in [6].

After running the simulation on a large number of parameter sets, we saw that in most cases, even in the presence of partitions, ITRA worked normally, as if there were no partitions; no situation of re-invocation of a client synched operation was generated. However, whenever this situation was generated, the partition was detected. The simulation shows that minor modifications in ITRA can make it partition proof, meaning that no data corruption occurs. An interesting task that is yet to be done is embedding our solution in ITRA and testing it in a real-life environment.

References

1. Dekel, E., Goft, G.: ITRA: Inter-Tier Relationship Architecture for End-to-end QoS. Journal of Supercomputing 28, 43–70 (2004)
2. Lamport, L.: Paxos Made Simple. ACM SIGACT News (Distributed Computing Column) 32(4), 51–58 (2001)
3. DESMO-J – Discrete-Event Simulation and Modeling in Java,
   http://asi-www.informatik.uni-hamburg.de/desmoj/
4. HP – Arbitration for Data Integrity in Serviceguard Clusters,
   http://docs.hp.com/en/B3936-90078/ch01.html
5. Microsoft – Windows Server 2003 Deployment Whitepapers - Server Clusters: Majority Node Set Quorum,
   http://technet.microsoft.com/en-us/library/cc784005.aspx
6. Aviv, D.: Handling Partitions in a Multi-tier Environment, Master Thesis, Haifa University (2009)

Short and Informal Documents: A Probabilistic Model for Description Enrichment

Yuval Merhav and Ophir Frieder

Information Retrieval Lab, Computer Science Department
Illinois Institute of Technology
Chicago, Illinois, U.S.A
yuval@ir.iit.edu, ophir@ir.iit.edu

Abstract. While lexical statistics of formal text play a central role in many statistical Natural Language Processing (NLP) and Information Retrieval (IR) tasks, there is little known about lexical statistics of informal and short documents. To learn the unique characteristics of informal text, we construct an N-gram study on P2P data, and present the insights, problems, and differences from formal text. Consequently, we apply a probabilistic model for detecting and correcting spelling problems (not necessarily errors) and propose an enrichment method that makes many P2P files better accessible to relevant user queries. Our enrichment results show an improvement in both recall and precision with only a slight increase in the collection size.

Keywords: N-grams, Spelling Problems, P2P, Descriptor Enrichment.

1 Introduction

Electronic text is generated daily by millions of users around the world. Among this information, we can find massive amounts of informal and short documents. These informal documents are not written for a broad audience, and writers do not obey strict grammatical conventions [13]. In addition, unlike formal documents, informal documents are usually very short (instant messages, media files descriptors, etc.). Since research in Information Retrieval (IR) primarily focuses on formal and self explanatory document search (i.e., newswire, literature, web pages, etc.), IR techniques tested on such documents are not necessarily advantageous for informal text.

For insight into the language conventions and characteristics of an informal and short document forum, we apply an N-gram study over the Peer-to-Peer (P2P) domain; we chose the P2P domain because P2P users generate informal and short documents, and real data that were crawled from the LimeWire's Gnutella system, using IR-Wire, a publicly available research tool [16], were available to us. In a P2P system, every user of the system has a set of shared files, and each file is described by a list of terms called a descriptor. Approximately 90% of the files in a P2P system are music, video, and image files [16], which unlike text files, are not self-descriptive. The short and informal file descriptor (including any metadata associated with it) is the main descriptive resource available to match user queries with files. Approximately 5.5 million P2P file descriptors were used in our study.

Y.A. Feldman, D. Kraft, and T. Kuflik (Eds.): NGITS 2009, LNCS 5831, pp. 109–120, 2009.
© Springer-Verlag Berlin Heidelberg 2009

Many P2P research efforts focus on the architectural design of such systems, trying to improve search methods in terms of cost, scalability issues, and fault tolerance. However, only a few P2P efforts address the problem of search accuracy for a given query, as compared to other systems such as web search engines. In search engines, the main focus is to find relevant documents as fast as possible, while in P2P systems, the file exchange between peers is the primary service. This might explain why P2P search methods are primitive as compared to the state of the art of web search.

Exploiting our N-gram study observations, we propose various approaches to improve search accuracy in a P2P system. Since many of the words/phrases contain different spelling problems that are more common to the P2P domain, we propose a probabilistic model that detects many of these problems and suggests alternatives. We show that our approach enables users to find files that exist on the network that they previously would not have been able to find.

2   N-Gram Statistics

2.1   Preprocessing

After removing major punctuation marks, irrelevant white space and symbols, we considered every file descriptor as a *sentence*, and extracted all **word** unigrams, bigrams, trigrams, four-grams and five-grams with their frequency counts. Since file descriptors are short and specific, no sentence parsing was used. Also, all letters were converted to lower case, and numbers were treated the same as words. In total, our processed data contained almost 28,000,000 tokens.

2.2   Unique N-Grams

Figure 1 shows the total number of unique N-grams; there are 524,797 unique tokens in our data (unigrams) with approximately an average frequency of 53 per token. Not surprisingly, the number of bigrams is significantly larger, with 2,299,585 unique bigrams. The number of unique N-grams does continue to increase from bigrams to trigrams, albeit at a lower rate of growth, but decreases for four-grams and five-grams. The reduction continues as N increases further. These results are different from a recent web N-gram study, where it was shown that the number of unique N-grams decreases only for N=6 and further [19]. This difference is due to the difference in size between web sentences and P2P file descriptors that are shorter on average and not grammatically structured. Furthermore, the factor by which the total number of N-grams increases from one N-gram to another is much smaller for P2P, which implies that on average, using statistics for predicting the next word given the previous words, would be more accurate for P2P than the web.

Since many P2P users represent their files in unique ways (e.g., emin3m g8est hits), we are also interested in the number of unique N-grams for those N-grams that appear a sufficient number of times. In Figure 2, we illustrate the number of unique N-grams that occur exactly once and that appear at least 10 times. Consequently,

N-grams that appear at least 10 times contribute a small percentage to the total number of unique N-grams.

N-grams with a frequency of one dominate the unique N-gram list. As shown in Table 1, except for bigrams, more than 60% of all N-grams appear only once. After manually examining a portion of these N-grams, it appears that the majority of them are either numbers or a combination of letters and numbers, or words that contain spelling problems (explained in greater detail in the next sections). File descriptors that include these low frequency N-grams do not contribute to the P2P network since it is unlikely that these files would be deemed relevant against user queries.

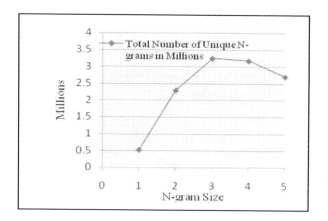

Fig. 1. Total number of unique N-grams, with N varying from 1 to 5. The y-axis is in millions.

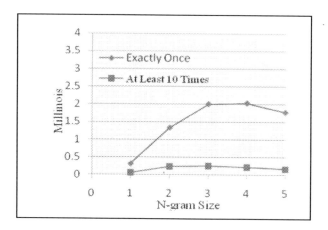

Fig. 2. Total number of unique N-grams, with N varying from 1 to 5, limited to those N-grams that appear at least 10 times, and those that appear only once. The y-axis is in millions.

Table 1. Percentage of N-grams that appear exactly once and those that appear at least 10 times, out of the complete list of unique N-grams

N-gram Size	Exactly Once	At Least 10 Times
Unigram	60.67%	12.12%
Bigram	58.16%	10.04%
Trigram	61.95%	7.79%
Four-gram	64.09%	6.79%
Five-gram	65.57%	6.18%

2.3 Most Frequent N-Grams

In Figures 3 and 4 we plotted the top 25,000 most frequent N-grams. In Figure 3(a) and 4(a), the plot is in log-log coordinates, while in 3(b) and 4(b), only the y-axis is in log scale. For all plots, the x-axis shows all top 25,000 most frequent N-grams, sorted from most to least frequent, and the y-axis shows the frequency of each N-gram. As expected, we see a few very frequent N-grams, and then a rapid decrease in frequency; this is similar to all N-grams, with unigrams having the most rapid decrease, as seen in the log-linear scale figures.

Another interesting observation from the log-log scale figures is that the points in each graph are close to a straight line, which implies that the frequency distribution of all N-grams, with N varying from 1 to 5, follows the well studied power law distribution (Zipf's law) [21]. In Figure 3(a), we can see a fitting of a straight line to the unigram's frequency curve. Although the fitting does not hold for the first few hundred most frequent unigrams, it fits nicely to the 'body' and 'tail' of the most frequent N-grams. Similarly, frequency distribution of N-grams of size 2 to 5, follow Zipf's power law. The reason Zipf's law does not hold for the first few hundred most frequent N-grams is because it assumes a faster frequency decrease than is actually found in our data.

It has been shown that for long and formal documents, the top ranked N-grams account for a very large portion of the total occurrences of all N-grams. For example, the top 10 ranked unigrams account for about 23% of the total token count [2]. In our data, the top 10 ranked unigrams account for about 10% of the total token count. This significant difference can explain the different role that stop words have in two distinct corpora like Brown [4] and our P2P data. Table 2 lists the percentage of the top 100, 1000 and 25,000 ranked N-grams occurrences out of all N-gram occurrences. It is interesting to see that the top 25,000 ranked unigrams (out of 524,797) account for 93% of the total token count; this means that about 500,000 (91%) tokens, only contribute 7% for the total token count. For N varying from 2 to 5, we can see a decreasing trend in the number of occurrences the top N-grams contribute to the total N-gram occurrences. For $N > 2$, the top 25,000 ranked N-grams account for less than 40% of the total N-gram occurrences, and the top 100 ranked N-grams account for less than 3% of total occurrences.

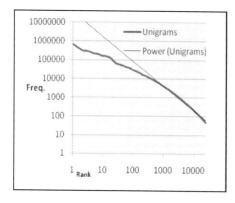
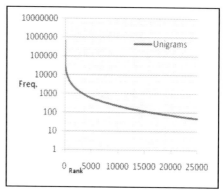

3(a) log-log coordinates   3(b) y-axis in log scale

Fig. 3. Frequency distribution of the 25,000 most frequent unigrams, ranked from most to least frequent

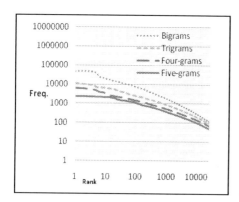
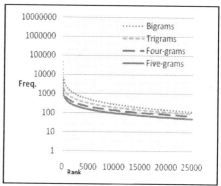

4(a) log-log coordinates   4(b) y-axis in log scale

Fig. 4. Frequency distribution of the 25,000 most frequent bigrams, trigrams, four-grams and five-grams, ranked from most to least frequent

Table 2. Percentage of top N-gram occurrences out of all N-gram occurrences, with N varying from 1 to 5

Rank	N-gram Size				
	N=1	N=2	N=3	N=4	N=5
Top 100	28.5%	5.7%	2.1%	1.5%	1.3%
Top 1000	59.9%	18.5%	8.6%	6.5%	6.1%
Top 25K	93.0%	55.7%	37.5%	31.8%	30.7%

3 Spelling Problems

Prior work on formal text spelling focuses mainly on typing errors caused by one of the following four transformations [14]:

1. Adding an extra letter
2. Deleting a letter
3. Replacing a letter
4. Transposing two adjacent letters

Experiments on formal text report that 80 to 90% of spelling errors are covered by the above transformations with edit distance 1 (i.e., only one transformation is needed to obtain the correct word) [14]. Based on our observations, we are interested in finding spelling problems which are not necessarily spelling errors. We consider every file descriptor to contain a spelling problem if any of its terms might make it difficult to be matched by user queries. Examples from our data include:

1. Typing/Spelling errors (e.g., 'eminen' instead of 'eminem'). Sometimes, a misspelled word is actually the correct use (e.g., the artist 'Krayzie Bone').
2. 'net speak' use by users and artists (some use 'in da club' (a correct song title), some 'in d club', some 'in the club', etc.)
3. Errors that are spelled correctly (e.g., 'boyz ii man' instead of 'boyz ii men'. Here although 'man' is not a misspelled word, it should be replaced with 'men'. Note that 'boyz' here is not a typo of 'boys'.) Many descriptors include 'to' or '2' instead of 'ii', and 'boys' instead of 'boyz'.
4. Adding unnecessary punctuation marks and numbers to words.
5. Writing two or more words as one word (e.g., 'nothing compares 2u').
6. Writing non English words using English letters (e.g., 'hevaynu sholem alaychem', a famous Hebrew song).

Our hypothesis is that P2P search accuracy is improved by rectifying the above problems. Since many of the above problems cannot be detected when ignoring the context (i.e., considering each word as a totally independent entity, which is how many traditional spelling methods work), we propose a bigram model for spelling problems in informal text.

3.1 The Bigram Model

At first, we use all word unigrams and bigrams with their frequency counts from file 1,000,000 descriptors as a training dictionary. Then, for each file descriptor D_i with terms $t_1, \ldots, t_n$ (n varies between descriptors) in our testing collection, we choose w_j which maximizes the probability of a term in position i:

$$\widehat{w}_j^i = \underset{w_j}{max}\, P(w_j) \cdot P(t_{i-1}, w_j) \cdot P(w_j, t_{i+1}) \tag{1}$$

$\widehat{w}_j^i$ is the estimation of the correct word w_j in position i. w_j is a word of distance one from t_i (which is not necessarily in our dictionary), including t_i itself. $P(w_j)$ is the probability to see w_j in our dictionary. $P(t_{i-1}, w_j)$ is the probability to see the bigram (t_{i-1}, w_j) in our dictionary; we can also refer to it as the joint probability of w_j and t_{i-1}. Note that t_{i-1} and t_{i+1} are the descriptor terms in positions $i-1$ and $i+1$, respectively. For the first position, we omit $P(t_{i-1}, w_j)$, and for the last position, we omit $P(w_j, t_{i+1})$. For unseen unigrams and bigrams, we use Laplace (add one) smoothing [10].

As an example, assume the short file descriptor $'in\ de\ club'$. To find the correct word in position two, we compute the following:

$$\widehat{w}_j^i = \underbrace{max}_{w_j} P(w_j) \cdot P('in', w_j) \cdot P(w_j, 'club')$$

The first w_j to consider is t_i (the original term $'de'$). Then, every possible candidate of edit distance one is considered as well. The w_j that maximizes the above equation is the one to be chosen as the correct term in position i (for most terms, it is the original term itself). Note that in the above example, the same process should be repeated at positions 1 and 3 as well. The corrected file descriptor will contain the terms which the model produced for each position, which is $'in\ da\ club'$. The intuition behind this model is that $P(w_j)$ helps to detect simple typing and spelling errors while $P(t_{i-1}, w_j)$ and $P(w_j, t_{i+1})$ can also detect more complicated spelling problems, like the ones previously described. Each term is assumed independent from the other terms, except the terms prior and next to it. Such an assumption is clearly false in reality, but term independence assumptions are common in search applications and rarely significantly impact accuracy.

4 Experiments

Our main goal is to show that solving different spelling problems improves search recall and precision. For our purpose, we used two separate collections of 1,000,000 file descriptors each, one used for training and the other for testing. We also used two randomly selected query sets (Q1 and Q2) of 30 queries each. The data sets used are publicly and freely available by request at http://ir.iit.edu/collections.

4.1 Testing the Model

To evaluate our model in terms of detecting and correcting spelling problems, we used 1200 randomly chosen descriptors; almost 30% of them contained at least one spelling problem. We applied our Bigram Model on each of the descriptors after training it on our training set. 69% of the spelling problems were detected and adjusted correctly, while 31% were not detected at all. For the correct descriptors, there were approximately 19% of false positive cases.

4.2 Descriptor Enrichment

Using our Bigram Model for detecting spelling problems, our aim is to enrich the descriptors; for each term in the descriptor that the Bigram Model suggested a replacement for, we added the suggested term to the descriptor without removing the original term; we refer to this method as Descriptor Enrichment (DE). Obviously, it is possible to enrich the descriptor by more than one term for each spelling problem, by adding the best k suggestions the Bigram model outputs. Adding more than one term may improve accuracy, but the tradeoff comes with an increased collection size. Another problem that may rise is that adding too many terms decreases search accuracy since some of these terms may be irrelevant to the original file. In our experiments, we compare between DE1, DE2, and DE3, which refer to descriptor enrichment by adding the top one, the top two, and the top three suggested terms, respectively. This means that for a given descriptor that contains t terms, the maximum number of terms that the new enriched descriptor can possibly contain is $t + t$, $t + 2t$, and $t + 3t$, for DE1, DE2, and DE3, respectively. In our experiments, the differences in the collection size before and after the enrichment were +14%, + 20%, and +33%, for DE1, DE2, and DE3, respectively (based solely on descriptors).

4.3 P2P Search

In today's P2P systems, the most popular search method is 'Conjunctive Queries' (CQ). In this method, for a given query, only file descriptors that match all query terms are deemed relevant. We use this method throughout all our experiments.

Figures 5(a) and 5(b) present the results obtained in our two search experiments (one for each query set). Each query was issued four times against the testing data; first run was made over the original collection, and the other three over the three enriched collections (DE1, DE2, and DE3). The enrichment method does not remove the original terms of the descriptors, and hence, every file that is part of the returned result set when issuing a query against the original collection, is also part of the returned result set when issuing the same query over each of the three enriched collections. However, the returned result set of any of the enrichment collections can include files that were not retrieved when the original collection was used; these files are the only ones that can contribute to the change in retrieval accuracy.

Consider the query *'jennifer lopez'*; an example of a file that was only acquired after applying search on one of the DE collections is *'selina jenifer lopez dreaming of you'*. This original descriptor did not contain all query terms, namely *'jennifer'*, and hence, could not be acquired using the given query over the original collection. After applying DE1, the descriptor was changed to *'selina selena jenifer jennifer lopez dreaming of you'* and was retrieved.

The number and distribution of all new files that were added to the returned result set, aggregated over 30 queries from Q1 and Q2, are presented in figure 5(a) and 5(b), respectively. For each of the three enrichment types, we evaluated the accuracy of the new files, labeling each file as either 'Same', 'Better', or 'Worse'. 'Same' is the group of files that were retrieved also when the original collection was used, but since the enrichment method changed their descriptor, they appear as new files. These files do not increase or decrease the search accuracy. 'Better' is the group of relevant files that could not be acquired using the original collection; the previous $'selina\ selena\ jenifer\ jennifer\ lopez\ dreaming\ of\ you'$ is an example of a file in this group. The files in this group are the ones that contribute to the increase in both precision and recall. Lastly, 'Worse' is the group of irrelevant files that were not acquired using the original collection, but were acquired using the enriched collection. The files in this group contribute to a decrease in precision. The files retrieved for each collection are the union of the files in the three groups 'Better', 'Same', and 'Worse'.

Our results show that all three enrichment methods improved the search accuracy. Since the 'Worse' group only contains a few files, the decrease in precision is insignificant. In contrast, the 'Better' group contains many files that their contribution to the overall precision is more significant. Table 6 summarizes the precision each enrichment method achieved; note that it is not the overall precision, only the precision of the new files that could not be acquired using the original collection.

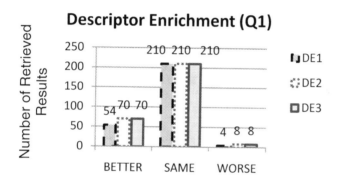

5(a) Query Set 1 (30 queries)

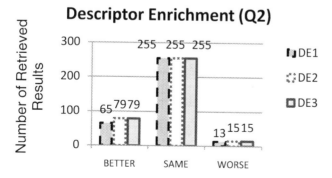

5(b) Query Set 2 (30 queries)

Fig. 5. The number of retrieved files added to the returned result list, obtained when using the DE1, DE2, and DE3 collections, but not when using the original collection

Table 6. Precision of the results presented in Figure 5

	Q1 Precision	Q2 Precision
DE1	93%	83%
DE2	90%	84%
DE3	90%	84%

In addition to the 83 to 93% precision performance for all the enrichment methods, there is also improvement in recall. This improvement is obvious since every file in the 'Better' group improves it. DE2 and DE3 retrieved the exact same files and share the exact same results, while DE1 achieved similar results with a smaller number of 'Worse' and 'Better' files. The difference between DE2 and DE1 is not statistically significant. We conclude that there is no need to enrich the descriptor with more than two terms for each problematic term, and both DE1 and DE2 are useful in making many files more accessible for user queries. Note that the precision improvement of DE1 is greater than that of DE2, but DE2 has a greater recall improvement. Hence, the possibility of using both enrichment options exists.

5   Related Work

The main reason of selecting the Bigram Model for detecting errors is its ability to detect context based spelling errors. Different methods to detect such errors have been proposed in the literature. Among the successful ones are Bayesian classifiers [5], latent semantic analysis [9], a combination of POS trigram based method and Bayesian methods [7], decision lists [20], Winnow-Based [6], and construction of confusion sets based on mixed trigram models [3].

Some prior efforts have focused on improving poorly described P2P file descriptors. When a peer in a P2P network downloads a file, it replicates its descriptor. The file is identified by a unique key, but its descriptor may vary between peers. Rich and detailed descriptors increase the odds to find relevant results. One proposed solution is to probe unpopular files, assuming those files are poorly described [8]. A peer gathers descriptor terms used by other peers that share the same file, and copies a few of those terms to its own file descriptor. Hopefully, those terms improve the file description, and increase the likelihood that the file is found. Terms can be picked using any criterion (e.g., randomly, term frequency, etc.). Identifying named entities has also been shown to be effective for improving precision in P2P search [11][12].

Other works on enrichment methods have focused on image retrieval. Images, like P2P files, contain an informal and short descriptor that is used to determine if the image is relevant for a given query (some systems also use computer vision techniques for image retrieval [1]). One popular method uses people to manually tag images while playing online games [17]; this way the expensive and boring task of manual tagging turns to be cheap and enjoyable to people who choose to play these games. This approach has the disadvantage of not being fully automated. A different approach suggests to start with a query image based on a keyword search, find similar images to it in terms of local or global features, extract phrases from their descriptions, and then select the most salient ones to annotate the query image [18]. The idea is that if you find one accurate keyword for a query image, then you can find complementary annotations to describe the details of this image. A similar approach for video annotation is described in [15].

6 Conclusions

The rising volume of short and informal content motivated us to apply an N-gram study for learning the characteristics of P2P data. We observed that a large portion of the descriptors contain spelling problems that make them less accessible to relevant queries. We applied a language independent probabilistic model for detecting spelling problems, and experimented enrichment methods that add terms to the problematic descriptors based on our probabilistic spelling corrector's output. Our experiments have shown that both precision and recall are improved when search is preformed on the enriched descriptors, instead of on the original collection. Our results confirm our hypothesis that spelling based enrichment of file descriptors improves recall and precision for many queries, with only a slight increase in the collection size.

References

1. Barnard, K., Forsyth, D.A.: Learning the Semantics of Words and Pictures. In: International Conference of Computer Vision, pp. 408–415 (2001)
2. Broni, M., Evert, S.: Counting Words: An Introduction to Lexical Statistics. In: 18th European Summer School in Logic, Language and Information, Spain (2006)
3. Fossati, D., Di Eugenio, B.: I saw TREE trees in the park: How to correct real word spelling mistakes. In: LREC 2008, 6th International Conference on Language Resources and Evaluation, Marrakech, Morocco (May 2008)
4. Francis, W.N., Kucera, H.: Brown Corpus Manual. Brown University, Providence (1964), http://www.hit.uib.no/icame/brown/bcm.html
5. Golding, A.R.: A Bayesian Hybrid Method for Context-Sensitive Spelling Correction. In: 3rd Workshop on very large corpora. ACL (1995)
6. Golding, A.R., Roth, D.: A Winnow Based Approach to Context-Sensitive Spelling Correction. Machine Learning 34(1-3), 107–130
7. Golding, A., Schabes, Y.: Combining Trigram-Based and Feature-Based Methods for Context Sensitive Spelling Correction. In: 34th Annual Meeting of the Association for Computational Linguistics, Santa Cruz, CA (1996)
8. Jia, D., Yee, W.G., Nguyen, L., Frieder, O.: Distributed, Automatic File Description Tuning in P2P File-Sharing Systems. Peer-to-Peer Networking and Applications 1(2) (September 2008)
9. Jones, M.P., Martin, J.H.: Contextual Spelling Correction using Latent Semantic Analysis. In: 5th Conference on Applied Natural Language Processing, Washington, DC (1997)

10. Jurafsky, D., Martin, J.H.: Speech and Language Processing: An Introduction to Natural Language Processing. In: Computational Linguistics and Speech Recognition. Prentice-Hall, Englewood Cliffs (2000)
11. Merhav, Y., Frieder, O.: On Filtering Irrelevant Results in Peer-to-Peer Search. In: 24th Annual ACM Symposium on Applied Computing (SAC 2008), Fortaleza, Brazil (2008)
12. Merhav, Y., Frieder, O.: On Multiword Entity Ranking in Peer-to-Peer Search. In: 31st Annual International ACM SIGIR, Singapore (2008)
13. Minkov, E., Wang, R.C., Cohen, W.W.: Extracting Personal Names from Email: Applying Named Entity Recognition to Informal Text. In: Conference on Human Language Technology and Empirical Methods in Natural Language Processing (2005)
14. Mitton, R.: English Spelling and the Computer. Longman, London (1996)
15. Moxley, E., Mei, T., Hua, X.S., Ma, W.Y., Manjunath, B.: Automatic Video Annotation through Search and Mining. In: IEEE International Conference on Multimedia and Expo., ICME (2008)
16. Sharma, S., Nguyen, L.T., Jia, D.: IR-Wire: A Research Tool for P2P Information Retrieval. In: 29th Annual International Workshop ACM SIGIR, Seattle, WA (2006)
17. Von Ahn, L., Dabbish, L.: Labeling Images with a Computer Game. In: Conference on Human Factors in Computing Systems, pp. 319–326 (2004)
18. Wang, X.J., Zhang, L., Jing, F., Ma, W.Y.: Annosearch: Image auto-annotation by search. In: IEEE Computer Society Conference on Computer Vision and Pattern Recognition. CVPR, pp. 1483–1490. IEEE Computer Society, Washington (2006)
19. Yang, S., Zhu, H., Apostoli, A., Pei, C.: N-gram Statistics in English and Chinese: Similarities and Differences. In: International Conference on Semantic Computing, USA (2007)
20. Yarowsky, D.: Decision Lists for Lexical Ambiguity Resolution: Application to Accent Restoration in Spanish and French. In: Annual Meeting of the Association for Computational Linguistics, Las Cruces, NM (1994)
21. Zipf. GK.: Relative Frequency as a Determinant of Phonetic Change. In: Reprinted from the Harvard Studies in Classical Philology, XL (1929)

Towards a Pan-European Learning Resource Exchange Infrastructure*

David Massart

European Schoolnet, rue de Trèves 61, B-1040 Brussels, Belgium

Abstract. The Learning Resource Exchange (LRE) is a new service that provides schools with access to educational content from many different origins. From a technical standpoint, it consists of an infrastructure that:

- Federates systems that provide learning resources – e.g., learning resource repositories, authoring tools – and
- Offers a seamless access to these resources by educational systems that enable their use – e.g., educational portals, virtual learning environments (VLEs).

As the number of connected systems increased over time, this infrastructure had to evolve in order to improve the quality of its search service.

This paper describes the current LRE infrastructure and explains the rationale behind its evolution.

1 Introduction

The Learning Resource Exchange (LRE) is a new service that enables schools to find digital educational content from many different countries and providers. It initially includes content from Ministries of Education (MoE) and other partners working with European Schoolnet (EUN) in large-scale European Commission-funded projects. Over 128,000 learning resources and assets were made available when the LRE as a public service started to be offered to schools and teachers on 1 December 2008. Additional resources from LRE Associate Partners are also being included in the LRE and the amount of content that schools can access is growing rapidly.

As suggested in Figure 1, the LRE provides access to learning resources from various origins. LRE content is provided by ministries of education (MoE), commercial and non-profit content providers (Publishers), and cultural heritage organizations (Museums). It might also include user-generated content (Teachers). Content is described with machine-readable descriptions, called metadata. Metadata is stored in repositories where it is exposed to the LRE that federates it.

---

* The work presented in this paper is partially supported by the European Community eContent*plus* programme - project ASPECT: Adopting Standards and Specifications for Educational Content (Grant agreement number ECP-2007-EDU-417008). The author is solely responsible for the content of this paper. It does not represent the opinion of the European Community and the European Community is not responsible for any use that might be made of information contained therein.

Y.A. Feldman, D. Kraft, and T. Kuflik (Eds.): NGITS 2009, LNCS 5831, pp. 121–132, 2009.

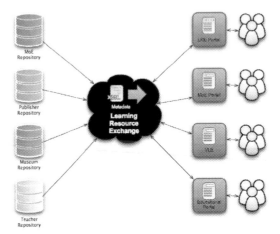

Fig. 1. An infrastructure for exchanging learning content

The LRE federation is not limited to a single access point. Potentially, any application that utilizes learning resources can connect to it including, for example, the LRE Portal (http://lreforschools.eun.org), a national portal hosted by a Ministry of Education (MoE Portal), a virtual learning environment (VLE) or any other third-party educational portal (Educational Portal).

Three main technical success factors have been identified for an infrastructure such as the LRE:

1. Semantic interoperability: The ability of two or more software systems to agree on the meaning of data or methods [1]. Semantic interoperability is key to enable LRE users to express relevant queries and to return query results that are relevant (precision) and complete (recall).
2. Reliability: The probability that a software system demonstrates failure-free performance under given conditions[1]. The LRE should be reliable enough so that teachers can count on it when in front of a class.
3. Efficiency: The ability of a software system to place as few demands as possible on hardware resources [2]. The LRE should be efficient enough to answer queries without an unreasonable delay.

In other words, the infrastructure should be reliable and efficient enough to provide its end-users (e.g., teachers, pupils) with access to learning resources adapted to their needs whenever they need them.

Early versions of the LRE were built on a brokerage system to which learning resource repositories, educational portals, and virtual learning environments connected to share learning resources [3]. Federated search was the main content discovery mechanism. Queries submitted by an educational portal end-user

---

[1] http://www.testability.com/Reference/Glossaries.aspx?
Glossary=Reliability - Last visited on Feb. 2, 2009.

were sent to all the metadata repositories connected to the LRE for references of learning resources matching the search criteria. In addition, metadata harvesting (i.e., mirroring) was proposed as an alternative way of exposing the metadata of a repository. Metadata stored in repositories that did not participate in the federated search was regularly mirrored in a repository that was directly connected to the federation [4].

An LRE metadata application profile [5] was used to unify the way learning resources were described in the federation. It was based on the IEEE Learning Object Metadata standard (LOM) [6] that was profiled (i.e., adapted) to the context of schools in Europe, enabling the LRE to offer good performances in terms of semantic interoperability.

This architecture was initially adopted because it was easy to implement by repositories while offering maximum flexibility: It was decentralized enough to allow content providers to manage their collections autonomously, and secure enough to ensure the trust needed when dealing with content for school pupils.

However, this model proved not to be scalable. It worked well only as long as the size of the federation did not exceed a dozen repositories. Beyond this, one observed a degradation of the infrastructure performance in terms of semantic interoperability, reliability, and efficiency.

This paper describes how the LRE technical infrastructure has evolved to accommodate a growing number of connected repositories and an increase in the volume of learning resource exchanged while improvements were made to the quality of its search service.

By enabling a federated metadata approach, this new infrastructure has particularly allowed various mechanisms to be implemented for:

– Caching metadata, which makes the search experience more stable (Section 2),
– Collecting information from resource providers (Section 3) and resource users (Section 5), which improves the search precision and recall, and
– Optimizing the resource description data model which improves the performance of search results display (Section 4),

2   From Federated Search to Federated Metadata

Federated search was the main content discovery approach deployed in the first versions of the LRE. Thanks to a search protocol such as the Simple Query Interface (SQI [7]), connecting a repository to the LRE federated search was easy and took typically a few days. It was also possible for repositories to use the Open Archive Initiative Protocol for Metadata Harvesting (OAI-PMH [8]) to expose their metadata for mirroring, but this protocol is more complex and its implementation usually takes longer. Initially, the main priority was to connect repositories and being able to offer fast and easy connection was key in order to convince content providers to join the federation.

Once the infrastructure was in place and enough repositories were connected, large-scale experiments undertaken in schools during the CALIBRATE (http://

calibrate.eun.org) and the MELT (http://info.melt-project.eu/) projects demonstrated that, when using the LRE, teachers' primary concerns were the stability and reliability of their search experience. When preparing a lesson at home, a teacher wants to be able to discover quality content adapted to her needs and to make sure that she will find this content again when the lesson is being delivered in front of a class.

The LRE repositories are heterogeneous. They were designed independently, belong to organizations with different technical skills and rely on very different technologies to manage descriptions of learning resources (e.g., relational databases, XML databases, object databases, file indexes, and other ad hoc solutions). As the number of participating repositories rose, it became clear that these repositories were not reliable enough to support a learning resource discovery service based solely on federated search.

There are two possible approaches to discover learning content in a federation of repositories such as the LRE:

1. Federated search, which consists of querying each repository of the federation separately and merging the different query results and
2. Federated metadata, which consists of gathering the metadata exposed by the repositories of the federation into a central cache and querying this cache.

Each approach has its pros and cons. During a federated search, repositories are queried in real time, whereas, with federated metadata, it is the cache that is queried. Real time searching of repositories always produces up-to-date results. This is an advantage when collections are volatile with frequent updates. Moreover, during a search operation, a repository can take advantage of unexposed information to process queries more efficiently (unexposed information is information about resources that a repository stores but does not make directly available either because it is not part of the selected metadata standard or because its access is restricted for some reason). As we have seen, one drawback of live searching is that repositories that are temporarily unavailable at query time, for whatever reason, are ignored. Moreover, this approach requires the merging of search results from different sources. In contrast, when searching cached metadata, complete results are always returned at once (even if some repositories are unavailable at the time of the query), but these results might be out of date in terms of the current state of play of repositories in the federation.

Taking into account the results of the first experiments in schools, it was decided to opt for the federated metadata approach to reach the level of service stability required by teachers.

3   Building Semantic Interoperability (Part I)

The ability to obtain quality information in a machine-readable format is key in order to achieve semantic interoperability. With the federated metadata approach, the only information available comes from the metadata exposed by the content providers. As the number of these providers increases (including partners that are not necessarily funded via projects to produce LRE metadata),

metadata quality – i.e., metadata completeness and accuracy with regard to the application profile – becomes problematic.

A way to improve metadata correctness and completeness involves working with the providers themselves by:

- Making them aware of potential errors in their metadata (cf. Section 3.1) and
- Taking advantage of their specificities to correct and complete their metadata (cf. Section 3.2).

3.1   Metadata Validation

Metadata is used to adequately describe learning resources. Thanks to these descriptions, resources become easier to find and teachers and pupils can more easily assess their usefulness.

In the LRE context, "adequately" means in a way adapted to the context of primary and secondary schools in Europe. The problem is three-fold:

- Primary and secondary schools have specificities in terms of organization, pedagogy, and curriculum.
- Although commonalities exist, these specificities vary from one European country (or region) to another.
- Multiple languages, which in Europe is the rule, not the exception.

The LRE metadata application profile [5] proposes a conceptual schema for describing learning resources in a way that is adapted to the European school sector. It profiles the IEEE 1484.12.1 Learning Object Metadata Standard (IEEE LOM) [6] as follows:

- It defines mandatory, recommended, and optional elements of the IEEE LOM standard data model. For example, LOM element 5.9 Educational Typical Age Range, (which, when relevant, is considered as the best way to refer to the audience of a resource regardless of the school system under consideration) is a recommended element.
- It introduces new controlled vocabularies, for example for LOM element "5.2 Learning Resource Type". Each new vocabulary is designed to take into account the specificities of primary and secondary education in Europe. In addition, each vocabulary is translated into different European languages including a neutral form that can be used as an 'inter-language' or bridge during the search and exchange of resource descriptions.

Ideally, all the LRE resources should be described according to the LRE metadata application profile. However, this is not always possible. Most legacy collections of learning resources have been tagged according to other application profiles (or have not been tagged at all) and the cost of retagging these resources is generally prohibitive. So, in practice, when a new content provider (or group of content providers) is interested in joining the LRE, negotiations and technical discussions start to explore to what extent the provider is able to support the

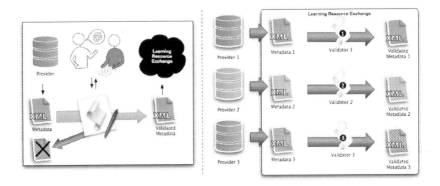

Fig. 2. A metadata validator is used to check that metadata complies with criteria agreed with its provider (left). Different validators are required to check metadata from different origins (right).

LRE metadata application profile. This process is illustrated by the left part of Figure 2. Once an agreement is reached, a technical solution or "validator" is configured to check that the provider's metadata complies with what was agreed. Non-compliant metadata instances are automatically rejected and a report is sent to the provider. In general, this reporting is a service appreciated by the providers as it enables them to detect problems they were not aware of and therefore helps them to improve the quality of their own metadata.

Metadata instances consist of XML documents. A validator goes beyond checking the validity of XML instances against the XSD binding of the LRE metadata application profile. It also allows for verifying the presence of mandatory elements, the use of controlled vocabularies, dependencies between elements, and broken URLs.

As suggested in the right part of Figure 2, multiple validators are necessary to enforce different agreements made with different (groups of) LRE providers.

3.2   Correcting and Completing Metadata

When new metadata records arrive in the LRE, they can be more or less complete and correct. They do not always comply with the LRE metadata application profile, which makes the resources that they describe much more difficult to find. In order to limit this problem, it is necessary to correct and complete the metadata as much as possible.

In some cases, information required by the LRE metadata application profile is present in the metadata but not expressed as required. For example, the target user of resources might be expressed in terms of grades (e.g., first grade, second grade) instead of age-range. If this information is expressed in a way that is systematic enough, it is relatively easy to write a script to automatically correct it.

Sometimes, some information missing in the metadata can be deduced from the context of a provider. For example, it is easy to complete metadata that does not contain license and learning resource type information when the resources of a given repository are all videos released under the terms of the same license.

4   Improving Efficiency: LRE Metadata Format

As already mentioned, the IEEE 1484.12.1 Learning Object Metadata (LOM) [6] is a standard format for transporting learning resource descriptions. It proposes a hierarchical conceptual data schema that denes the structure of metadata for a learning resource. An IEEE LOM instance consists of a tree structure of data elements that describe different characteristics of a learning resource and are grouped in general, life cycle, meta-metadata, educational, technical, educational, rights, relation, annotation, and classication categories. Some of these data elements, for example, the title of a learning resource, are of a special data type named "language string". This data type allows for multiple semantically equivalent character strings, for example, the translations into different languages of the title.

Usually, a limited number of LOM elements are displayed in search results, ideally in the language of the end-user. Typically, these elements are the title, description, learning resource type, location, and possibly some other information about the resource. Since this information comes from different LOM categories and language strings, extracting it requires the parsing of the entire LOM structure, which can be an expensive and time-consuming overhead for queries that returns many results.

To overcome this limitation, after having been completed and corrected as described in Section 3.2, LOM instances are turned into an optimized data structure called a Cached Metadata Record (CMR). As depicted in Figure 3, a CMR consists of 3 parts: Identifiers, language blocks, and indexes.

The identifiers part proposes identifiers to uniquely identify the learning resource described according to different schemes: The original LOM identifier is given a numeric identifier that is managed internally by the LRE.

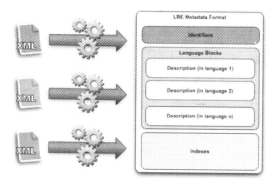

Fig. 3. Turning LOM records into LRE cached metadata records (CMR)

A language block is a flat structure that only contains the subset of LOM information in a given language that is necessary to display a search result item. Typical information for a language block includes the title, description, keywords, and target age range of the resource, the url where it can be accessed and a pointer to its full metadata record. When a CMR is received as a search result, the processing required is limited to selecting the block in the language of the end-user and to displaying it. This approach proved to be much faster than extracting the same information from a LOM record.

Finally, the third part contains all the information that needs to be indexed.

5   Building Semantic Interoperability (Part II)

5.1   Metadata Enrichment

Content providers are only one possible source of information about learning resources. Other sources of information that can be used to enrich learning resource descriptions include: Users of these resources, controlled vocabularies used in the metadata, professional indexers, and the resources themselves.

The development and use of automatic metadata generation tools (i.e., tools that produce metadata from learning resources) are still experimental in the LRE and will not be described here as such tools have not been integrated yet in the production workflow of the federation.

The way teachers and pupils use learning resources provides useful information about these resources. Figure 4 shows a screenshot of one of the portals connected to the LRE (`http://lreforschools.eun.org`) where teachers have the possibility to add their own tags (i.e., free keywords) to resources (4), to rate resources (2) by attributing them from 0 (very bad) to 5 stars (outstanding), and to bookmark their favorite resources (3). In addition, a "travel well" factor is evaluated for each resource based on the number of times this resource has been used in a cultural and linguistic context different from the one in which it has been created (1).

The different LRE portals are encouraged to keep track of this information generated by their users' activities and to expose it in a standard way so that

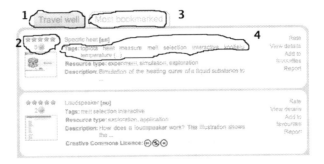

Fig. 4. Screenshot of the LRE for Schools portal

it can be easily collected by the LRE. Then it is consolidated, and added to resource descriptions where it is used to enrich indexes (with user-generated tags) and improve the sorting of search results (by rating, most bookmarked, and travel).

Currently available in 15 languages, the LRE Thesaurus is a family of 25 micro thesauri produced to describe aspects related to learning resources such as "Content of education", "Educational system", "School activities"[2]. The LRE takes advantage of the thesaurus relationships to add to the description of a resource tagged with a thesaurus descriptor (e.g., "domestic animal") all the broader terms of this descriptor (e.g., "animal", "zoology", "biology", "biological sciences", "natural sciences", and "sciences") so that a search on one of these broader terms (e.g., biology) will return resources tagged with a more specialized descriptor (narrower term).

Another way of enriching the LRE metadata consists of asking professional indexers to manually tag resources according to the LRE Metadata Application Profile. This is usually reserved for LRE resources belonging to collections considered as having a high educational and "travel well" potential.

5.2 Automatic Metadata Translation

Translating metadata is another way of improving the discoverability of learning resource. From a translation standpoint, one can distinguish between three types of metadata information: Identifiers, controlled vocabulary entries, and free texts.

Identifiers are generally unique sequences of characters with no meaning. They are used to identify learning resources and metadata. As they are language-independent, they do not need to be translated.

Controlled vocabulary entries are easy to translate. For all the controlled vocabularies of the LRE Metadata Application Profile, each vocabulary entry consists of:

− A token used as neutral representation of a given concept and
− An approved translation of this concept in each of the supported languages.

Using the LRE Metadata Application Profile XML binding, only neutral tokens are stored in the metadata. Automatically obtaining the translation of such a token into a given language is very simple. It consists of looking up in a vocabulary database for the token translation in the requested language. Neutral tokens are stored in the index part of CMRs. Their translations in a given language are stored in the CMR language block for this language.

Each CMR language block is divided into two subparts: One for controlled vocabulary entry translation, the other for free text metadata elements such as title, description and free keywords. The latter can be easily extracted and passed

---

[2] More information on the LRE Thesaurus is available at
http://lre.eun.org/node/6

to an automatic translation service[3]. Although the quality of these automatic translations can vary a lot, experience shows that, once these automatically translated texts have been tokenized and stop-words filtered out, the resulting keywords are generally good enough to usefully contribute to the index part of CMRs.

These automatic translations are also stored in language blocks with special language identifiers that clearly identifies them as machine translations[4]. These machine-translated language blocks are used in two ways.

- When a human translator wants to translate a block into a language for which a machine translation exists, the latter is used to pre-fill the translation form.
- When a user finds a resource for which no description exists in her language, she is informed that she has the possibility to request a machine translation for it.

5.3 Metadata Indexing

Once a CMR has been enriched, it can be indexed. Currently, the LRE supports the following indexes:

- Free text: All the free text metadata elements used in a resource description (title, description, keywords), including their human and automatic translations; the translation of the controlled vocabulary entries; and the tags provided by end-users.
- Descriptor: The token forms of the LRE Thesaurus entries used to index a resource and the token form of their broader terms.
- Learning resource type: The token form of the learning resource type vocabulary entries used to describe a resource.
- Age range: The age range of the target audience of the learning resource.
- Rights: A machine-readable representation of the license that applies to the resource.
- Resource language: The language(s) of the learning resource.
- Metadata language: The language(s) in which the learning resource description is available.
- Metadata provider: The identity of the organization that provided the metadata of the resource.

In addition, the "travel well" factor of a resource, the number of times it has been bookmarked, and its rating (as attributed by users) are used to sort query results but are not searchable yet.

---

[3] The current version of this LRE service is based on Systran and supports 8 language pairs: from English to French, Italian, Dutch, German, Spanish, Portuguese, Greek and Russian.

[4] This code consists of prefixing the two-letter ISO code of the target language (e.g., "en") with the prefix "x-mt-" (e.g., "x-mt-en").

5.4   Learning Resource Identity and Metadata Management

The LRE does not handle learning resources directly. It manages learning resource metadata that is subject to change. Learning resource descriptions are updated, either because the learning resources that they describes evolve or simply because new information about these resources become available. Moreover, different LRE providers can potentially have different descriptions of the same learning resource.

In this context, each time a new metadata record is acquired by the LRE, it is important to clearly identify the resource that is described.

The LRE maintains a list of resource identifiers. Each time a new learning resource description arrives in the LRE, it is compared to existing descriptions to see if it refers to a new learning resource or to a learning resource for which a description already exists in the LRE. This comparison is not trivial since, although a part of a new description is generally unchanged compared to an old one, any element of this description can potentially have changed, including the identifiers and locations of the resource.

Once the new metadata record is recognized as describing a resource unknown by the LRE, it receives a new identifier, is enriched, and a new CMR is generated. If a description already exists for this resource, the two descriptions are compared again and, if needed, the existing CMR is updated and possibly re-enriched.

The management of the learning resource identifiers is a key function of the LRE since the applications that utilize the LRE resources rely on them to manage their users' tags, rating, bookmarks, etc. and to feed this information back to the LRE.

6   Conclusion

This paper describes an improved technical infrastructure for exchanging learning resources that allows for enriching learning resource metadata before caching them in an optimized data structure. These improvements have enabled European Schoolnet to migrate the Learning Resource Exchange from being a successful proof-of-concept prototype into a production environment to which educational portals (such as the LRE for schools) and national education portals (such as Scoilnet) can connect in order to have access to hundreds of thousands learning resources/assets from more than 20 repositories.

At present, there is no way for the LRE to automatically discover new repositories. Currently, the locations of the LRE repositories and the description of the protocols they support for exposing their resources have to be entered manually. Future research directions include the development of an active registry that can be queried to find repositories hosting relevant resources and that will facilitate connections to them in an automatic way.

In early 2009, the content exchanged in the LRE federation consists of open educational resources (i.e., public domain content or copyrighted materials licensed under an open license such as Creative Commons or the GNU Free Documentation License). Other developments are now being explored that aim to support more complex distribution models. Examples of such models include:

- License-based access: In this model, buying learning content is similar to buying software. An individual user, a group of users or all end users of a LRE portal can access a learning resource or a group of learning resources when the license is granted.
- Credit-based access: In this model, end-users 'purchase' access to digital resources by spending LRE credits. Basically, users have a certain amount of LRE credits that they may choose to spend in order to gain access to LRE content. This situation is similar to that where an end-user would go to a shop to buy some goods with real money.

References

1. Ternier, S., Massart, D., Van Assche, F., Smith, N., Simon, B., Duval, E.: A simple publishing interface for learning object repositories. In: Proceedings of World Conference on Educational Multimedia, Hypermedia and Telecommunications 2008, Vienna, Austria, AACE, June 2008, pp. 1840–1845 (2008)
2. Meyer, B.: Object-Oriented Software Construction, 2nd edn. Prentice Hall, Englewood Cliffs (1997)
3. Colin, J.N., Massart, D.: LIMBS: Open source, open standards, and open content to foster learning resource exchanges. In: Kinshuk, R.K., Kommers, P., Kirschner, P., Sampson, D., Didderen, W. (eds.) Proc. of The Sixth IEEE International Conference on Advanced Learning Technologies, ICALT 2006, Kerkrade, The Netherlands, July 5 - 7, pp. 682–686. IEEE Computer Society, Los Alamitos (2006)
4. Massart, D.: The EUN Learning Resource Exchange (LRE). In: Chang, B., Kashihara, A., Kay, J., Lee, J., Matsui, T., Okamoto, R., Suthers, D., Yu, F.Y. (eds.) The 15th International Conference on Computers in Education (ICCE 2007) Supplementary Proceedings, vol. 1, pp. 170–174 (2007)
5. Van Assche, F., Massart, D. (eds.): The EUN Learning Resource Exchange LOM Application Profile, Version 3.0 (June 2007)
6. IEEE Standards Department: IEEE 1484.12.1-2002, Learning Object Metadata Standard (July 2002),
   http://ltsc.ieee.org/wg12/files/LOM_1484_12_1_v1_Final_Draft.pdf
7. Simon, B., Massart, D., Van Assche, F., Ternier, S., Duval, E.: A simple query interface specification for learning repositories. In: CEN Workshop Agreement (CWA 15454) (November 2005), ftp://ftp.cenorm.be/PUBLIC/CWAs/e-Europe/WS-LT/CWA15454-00-2005-Nov.pdf
8. Lagoze, C., Van de Sompel, H., Nelson, M., Warner, S.: The open archives initiative protocol for metadata harvesting version 2.0. Document Version 2004/10/12T15:31:00Z (2002),
   http://www.openarchives.org/OAI/2.0/openarchivesprotocol.htm

Performance Improvement of Fault Tolerant CORBA Based Intelligent Transportation Systems (ITS) with an Autonomous Agent

Woonsuk Suh[1], Soo Young Lee[2], and Eunseok Lee[3]

[1] National Information Society Agency
NIA Bldg, 77, Mugyo-dong Jung-ku Seoul, 100-775, Korea
sws@nia.or.kr
[2] Korea Automotive Technology Institute, 74, Yongjung-Ri, Pungse-Myun, Chonan,
Chungnam, 330-912, Korea
sylee@katech.re.kr
[3] School of Information and Communication Engineering, Sungkyunkwan University
300 Chunchun Jangahn Suwon, 440-746, Korea
eslee@ece.skku.ac.kr

Abstract. The ITS is a state-of-the-art system, which maximizes mobility, safety, and usefulness through combining existing transport systems with information, communication, computer, and control technologies. The core functions of the ITS are collection, management, and provision of real time transport information, and it can be deployed based on the Common Object Request Broker Architecture (CORBA) of the Object Management Group (OMG) efficiently because it consists of many interconnected heterogeneous systems. Fault Tolerant CORBA (FT-CORBA) supports real time requirement of transport information stably through redundancy by replication of server objects. However, object replication, management, and related protocols of FT-CORBA require extra system CPU and memory resources, and can degrade the system performance both locally and as a whole. This paper proposes an improved architecture to enhance performance of FT-CORBA based ITS by generating and managing object replicas autonomously and dynamically during system operation with an autonomous agent. The proposed architecture is expected to be applicable to other FT-CORBA based systems.

Keywords: Autonomous Agent, FT-CORBA, ITS, Object.

1 Introduction

The application of advanced communications, electronics, and information technologies to improve the efficiency, safety, and reliability of transportation systems is commonly referred to as intelligent transportation systems (ITS). The key component of ITS is information systems to provide transport information in real time which have characteristics as follows. First, these systems run on nationwide communication networks because travelers pass through many regions to reach their destinations.

Y.A. Feldman, D. Kraft, and T. Kuflik (Eds.): NGITS 2009, LNCS 5831, pp. 133–145, 2009.
© Springer-Verlag Berlin Heidelberg 2009

Second, travelers should be able to receive real time information from many service providers, while driving at high speed and transport information should be able to be collected and transmitted to them in real time. Third, the update cycle of transport information to travelers is 5 minutes [20]. As a result, the ITS is characterized by various service providers and a stable service environment.

The ITS is deployed by various independent organizations and therefore is operated on heterogeneous platforms to satisfy the characteristics, functions, and performance requirements described earlier. Accordingly, it is appropriate to build ITS based on the CORBA. However, FT-CORBA with stateful failover is needed to satisfy real time requirements of transport information considering the update cycle of 5 minutes. In stateful failover, checkpointed state information is periodically sent to the standby object so that when the object crashes, the checkpointed information can help the standby object to restart the process from there [18]. FT-CORBA requires additional system CPU and memory resources and therefore an architecture is required to minimize resource usage and prevent performance degradation of both local subsystems as well as the entire system due to fault tolerance protocols. Accordingly, this paper proposes an agent based architecture to enhance the performance of FT-CORBA based ITS. Due to the real time and composite characteristics of ITS, the proposed architecture is expected to be applicable to most applications. The remainder of this paper is organized as follows. In section 2, CORBA based ITS and FT-CORBA related work are presented. In section 3, the proposed architecture introduces an autonomous agent to enhance performance of FT-CORBA based ITS. In section 4, the performance of the proposed architecture is evaluated by simulation focused on usage of CPU and memory. In section 5, this research is concluded and future research directions are presented.

2   Related Work

There are several representative CORBA based ITS worldwide. In order to efficiently manage the planned road infrastructure and estimated increase in traffic volume by 2008, as well as the traffic generated during the 2008 Olympic Games, the Beijing Traffic Management Bureau (BTMB) in China had built an ITS using IONA's Orbix 2000 [10]. The Los Angeles County in US coordinates multiple traffic control systems (TCSs) on its arterial streets using a new Information Exchange Network (IEN) whose network backbone is CORBA software [2]. The Dublin City Council in Ireland has selected IONA Orbix™ as the integration technology for an intelligent traffic management system [10]. The Land Transport Authority in Singapore performed the 'traffic.smart' project, which is based on CORBA [9]. The Incheon International Airport Corporation in Korea had built information systems including ITS based on IONA Orbix 2.0 [11].

The Object Management Group (OMG) established the FT-CORBA which enhances fault tolerance by creating replicas of objects in information systems based on the CORBA. The standard for FT-CORBA aims to provide robust support for applications that require a high level of reliability, including applications that require more reliability than can be provided by a single backup server. The standard requires that there shall be no single point of failure. Fault tolerance depends on entity redundancy, fault detection, and recovery. The entity redundancy by which this specification provides fault tolerance is the replication of objects as mentioned earlier. This strategy allows greater flexibility in configuration management of the number of replicas, and of their assignment to different hosts, compared to server replication [17]. The requirements for replication styles of the FT-CORBA standard are as follows [8] [17]. 1) COLD PASSIVE – In the COLD PASSIVE replication style, the replica group contains a single primary replica that responds to client messages. If the primary fails, then a backup replica is spawned on-demand and the state of the failed primary is loaded into that replica, which then becomes the new primary. 2) WARM PASSIVE – In the WARM PASSIVE replication style, the replica group contains a single primary replica that responds to client messages. In addition, one or more backup replicas are pre-spawned to handle crash failures. If a primary fails, a backup replica is selected to function as the new primary and a new backup is created to maintain the replica group size above a threshold. The state of the primary is periodically loaded into the backup replicas, so that only a (hopefully minor) update to that state will be needed for failover. 3) ACTIVE – In the ACTIVE replication style, all replicas are primary and handle client requests independently of each other. FT-CORBA uses reliable multicast group communication [3, 5, 12, 19] to provide ordered delivery of messages and to maintain state consistency among all replicas. The infrastructure sends a single reply to the client by detecting and suppressing duplicate replies generated by multiple members of the object group.

End-to-end temporal predictability of the application's behavior can be provided by existing real-time fault tolerant CORBA works such as MEAD and FLARe [1][14]. However, they also adopt replication styles of FT-CORBA mentioned earlier as they are. Active and passive replications are two approaches for building fault-tolerant distributed systems [6]. Prior research has shown that passive replication and its variants are more effective for distributed real time systems because of its low execution overhead [1]. The WARM_PASSIVE replication style is considered appropriate in ITS in terms of service requirements and computing resource utilization. In practice, most production applications use the WARM PASSIVE replication scheme for fault tolerance. It is recommended in the field of logistics according to FT-CORBA specification as well. However, a method is required to maintain a constant replica group size efficiently.

FT-CORBA protocols need additional CORBA objects such as the Replication Manager and Fault Detectors, server object replicas, and communications for fault tolerance, and therefore require accompanying CPU and memory uses, which can cause processing delays, thereby deteriorating the performance. Processing delay can be a failure for real time services of transportation information. Natarajan et al. [16] have studied a solution to dynamically configure the appropriate replication style, monitoring style of object replicas, polling intervals and membership style. However, a method to maintain minimum number of replicas dynamically and autonomously, which means adjusting "a threshold" specified in the warm passive replication style for resource efficiency and overhead reduction of overall system, has not been reported.

3 Proposed Architecture

The FT-CORBA can be represented as Fig. 1 when an application uses the WARM PASSIVE style.

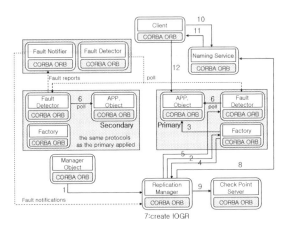

Fig. 1. FT-CORBA Protocol

The processes of Fig. 1 are summarized as follows [15]. 1. An application manager can request the Replication Manager to create a replica group using the create object operation of the FT-CORBA's Generic Factory interface and passing to it a set of fault tolerance properties for the replica group. 2. The Replication Manager, as mandated by the FT-CORBA standard, delegates the task of creating individual replicas to local factory objects based on the Object Location property. 3. The local factories create objects. 4. The local factories return individual object references (IORs) of created objects to the Replication Manager. 5. The Replication Manager informs Fault Detectors to start monitoring the replicas. 6. Fault Detectors polls objects periodically. 7. The Replication Manager collects all the IORs of the individual replicas, creates an Interoperable Object Group References (IOGRs) for the group, and designates one of the replicas as a primary. 8. The Replication Manager registers the IOGR with the Naming Service, which publishes it to other CORBA applications and services. 9. The Replication Manager checkpoints the IOGR and other state. 10. A client interested in the service contacts the Naming Service. 11. The Naming Service responds with the IOGR. 12. Finally, the client makes a request and the client ORB ensures that the request is sent to the primary replica. The Fault Detector, Application Object, and Generic Factory in Fig. 1 are located on the same server.

The administrator of ITS can manage numbers of object replicas with the application manager in Fig. 1 by adjusting fault tolerance properties dynamically. However, administration of ITS needs to be performed autonomously and dynamically with minimal intervention by the administrator. In addition, the use of system CPU and memory resources in FT-CORBA is large, which can affect the real time characteristics of ITS due to processing delays because FT-CORBA is an architecture to enhance fault tolerance based on the redundancy of objects. Accordingly, it is possible to enhance efficiency and prevent potential service delays if an autonomous agent (FTAgent) is introduced to the FT-CORBA based ITS, which adjusts the minimum numbers of object replicas autonomously and dynamically. It can be applied to other applications based on FT-CORBA. An autonomous agent is a system situated within and a part of an environment that senses that environment and acts on it, over time, in pursuit of its own agenda, and so as to effect what it senses in the future [7]. The FTAgent has algorithm and database [13] which can help to maintain the number of replicas efficiently because they require system CPU and memory resources both directly and indirectly, which can lower performance in terms of the overall ITS as mentioned earlier. The FTAgent is introduced in Fig. 2 on the same system as the Replication Manager in Fig. 1 which maintains 3 replicas for each object in this paper, i.e., the primary, first secondary, and second secondary replicas.

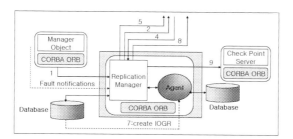

Fig. 2. Architecture to improve FT-CORBA

The FTAgent maintains its DB to support the Replication Manager for management of object replicas whose schema is as shown in Table 1.

Table 1. DB maintained by the FTAgent

IOGR IDs	date(dd/mm/yy)	time	failure 1	failure 2	flag	$risky_k$
1	01/01/08	00:00:00~00:04:59	0	0	0	0
1	.	00:05:00~00:09:59	0	0	0	0
1	.	.	.	.	.	.
1	.	23:50:00~23:54:59	1	1	1	10
1	01/01/08	23:55:00~23:59:59	1	1	1	11
1	02/01/08	00:00:00~00:04:59	1	0	0	0
1	.	.	.	.	.	.
1	31/01/08	23:55:00~23:59:59	0	1	0	0
.	.	.	.	.	.	.
100	31/01/08	23:55:00~23:59:59	0	1	0	0

The IOGR IDs identify replica groups of each object whose numbers are 100 in this paper. The numbers of records in Table 1 are maintained to be under 1 million because values of the time attribute of Table 1 are measured by 5 minutes per day. The date identifies days of one month. The time is measured every 5 minutes. The failure 1 means failures of primary object replicas which are original or recovered from previous failures. The failure 2 means failures of first secondary replicas after becoming the primary ones. The values of failure 1 and 2 are 0 for working and 1 for failed, respectively. The flag has two values which are 0 when primary or first secondary is working and 1 when both primary and first secondary have failed for respective 5 minutes which are represented as a service interval in Fig. 3. The $risky_k$ is a fault possibility index for object groups, which is assigned to each interval of 5 minutes for one hour backward from current time, and is set to zero at first. The k and $risky_k$ are equivalent and they ranges from 0 to 11 because the flag is set to 1 up to a maximum of 11 times for one hour. The values are assigned in the way that 11 and 0 are assigned to the nearest and furthest intervals of 5 minutes to current time, respectively.

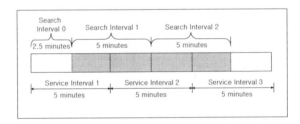

Fig. 3. Service interval of transport information and search interval for fault tolerance

The FTAgent searches the DB managed by Replication Manager and updates states (failed or working) of primary and first secondary replicas of each object (1~100) on its own DB in real time resuming every 5 minutes which ranges from previous to next middles of the information service interval of 5 minutes, restricted to one month (last 30 days) from current time. Search intervals are set between the respective middles of the former and latter service intervals because the moment of updating transport information is more important than any other time.

The FTAgent identifies whether there are simultaneous failures of primary and first secondary replicas of each object by searching its DB in real time. Object faults of ITS result from recent short causes rather than old long ones because it is influenced by road situations, weather, and traffic, etc., which vary in real time. If simultaneous failures for 5 minutes have originated for one month until now that the first secondary replica crashes which has been promoted to the primary as soon as the original primary one has failed and it is in the rush hours, the FTAgent requires the Replication Manager to keep the number of replicas of related objects 3, otherwise to reduce it to 2. In other words, the FTAgent lets the Replications Manager adjust the number of object replicas autonomously and dynamically. The decision by the value of the

parameter rush hours of whether it is in the rush hours is beyond this paper and depends on judgment in terms of traffic engineering. The algorithm of the FTAgent is
described as follows.

```
FTAgent(int rush hours){
  while(there is no termination condition){
(1)  search whether primary replicas of each object are
     working on the DB maintained by Replication Manager
     (RM) in real time resuming every 5 minutes which
     ranges from previous to next middles of the informa-
     tion service interval of 5 minutes, restricted to
     last 30 days from current time;
(2)  if(primary replica is working){failure 1=0 for all
        object groups identified by IOGRs; flag=0;}
(3)  else{failure 1=1 for all object groups;
(4)      confirm whether first secondary of each object pro-
         moted to primary by RM is working on the RM DB;
(5)      if(first secondary is working){failure 2=0;flag=0;}
(6)      else{failure 2=1;
(7)          confirm whether the replica created by RM,
             substituting for crashed primary is working;
(8)          if(it is working){failure 1=0; flag=0;}
(9)          else flag = 1;}}
(10)Decision_Number_of_Replicas(rush hours);}}

Decision_Number_of_Replicas(int rush hours){
(11)an array for numbers of two successive 1's of flag
    values for all object groups=0;
(12)search successions of two 1's in flag values for all
    object groups;
(13)if(there are two successive 1's of flag values) add
    to the number of two successive 1's of flag values
    for relevant objects;
(14)if{(number of two successive 1's ≥ 1 for last one
       hour)and(rush hours)}{
```

(15) $NoROR = [3-3 \times \{max(risky_k)/11\}]/3; NoROR_1 = NoROR;$

(16)   if$(0 \leq k \leq 5)\{NoROR = \{\sum_{d=1}^{30}(d \times NoROR_d)\}/30/30; NoROR_2 = NoROR;\}$

```
(17)    select the smaller one between NoROR₁ and NoROR₂,
        round it off, and assign the result to NoROR;
(18)    let RM keep the number of relevant object replicas
        minus NoROR, whose selection is the order of their
        ID numbers;}
(19)else let RM reduce the number to 2 which mean the two
        of the 3 replicas which are working at the moment
        and whose priority for selection is the order of
        their ID numbers;}
```

In line (15), NoROR stands for the number of reduced object replicas and in line (16), $NoROR_d$ means the number of reduced object replicas in the same time zones of 5 minutes for last 30 days. The proposed architecture in this paper can be applied to the work such as MEAD and FLARe to increase resource availability and decrease overheads by enhancing utilization efficiency of CPU and memory, thereby improving end-to-end temporal predictability of the overall system.

4 Evaluations

The items for performance evaluation are total time of CPU use and maximum usage of memory of servers related to the 11 processes except for the 12th process in Fig. 1 from the beginning to termination of the simulation of two types to maintain 3 and 2 object replicas for fault tolerance [4]. The simulation has been performed on the PC with Intel Pentium M Processor 1.60 GHz, 1.0 GB memory, and Windows XP as the OS. The programs are implemented in Visual C++ 6.0. Firstly, the use rate of CPU during simulation is 100% on the implementation environment, and therefore it is appropriate to measure and compare total times of CPU use from beginning to termination of the simulation programs of two types. They must be measured for all servers related to creation and maintenance of object replicas in Fig. 1. The processes without numbers on arrows in Fig. 1 are not considered. Accordingly, the number of CPUs to be considered is 11. The total time of CPU use is expressed as shown in formula (1).

$$\text{Total Time of CPU Use} = \sum_{i=1}^{m} \text{CPU-time}_i$$

(1)

CPU-time $_i$: Time of CPU use of each server,
m : numbers of CPU considered for simulation = 11

Secondly, the peak usage is more appropriate for memory rather than continuous measurement of memory use. Therefore, the maximum usage of two types of 3 and 2 replicas is measured respectively. Total time of CPU use and maximum usage of memory are compared in that the Replication Manager maintains 3 and 2 replicas of objects respectively. Namely, the 11 processes prior to the client requesting services in Fig. 1 are simulated with 2 separate programs which describe the two types in terms of CPU and memory use. The components of the FT-CORBA are the same and therefore they are not designed in the programs in terms of memory use. The processing latencies with loops in the programs are set for the type of 3 replicas as follows: 1) latency between internal components: 2 sec. 2) latency between external components: 3 sec. 3) latency for the FTAgent to search the DB maintained by the Replication Manager and itself and to deliver related information to it : 5 sec. Of course, latencies directly related to creating and maintaining 2 replicas are set to two thirds of those for 3 replicas. The values of the established processing latencies are variable due to basic processes of the OS in the implementation environment, which is ignored because the variableness originates uniformly in simulations of both types to be compared. The conditions presented earlier are based on the experimental fact that the processing latency to select records which have the condition of the line (14) in the algorithm is about 3 seconds in case of the Oracle 9i DBMS which maintains 1 million records with 13 columns on IBM P650 with 4 CPUs of 1.24GHz and 12GB memory, and is 34 Km distant from a client.

A commercial internet browser is used as an object to simulate usage of CPU and memory in creation and termination of 3 and 2 object replicas obviously. The browser is called 3 and 2 times by types and kept as processes until the simulation is finished. The items to be compared are total time of CPU use and maximum usage of memory from the beginning to termination of the simulation as mentioned earlier. The types of 3 and 2 replicas are simulated respectively by executing the relevant programs 5 times where www.sun.com outside Korea is filled in the URL of the browser assumed as an object. The results for the total CPU time used are shown in Fig. 4.

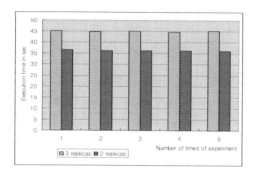

Fig. 4. Total time of CPU use in sec.

The total time of CPU use ranges from 44.69 to 45.20 seconds for the type of 3 replicas. The arithmetic mean is 44.97 seconds and the standard deviation is 0.23 seconds, which is 0.5% based on the minimum of 44.69 seconds. On the other hand, the total time of CPU use ranges from 36.13 to 36.59 seconds for the type of 2 replicas. The arithmetic mean is 36.29 seconds and the standard deviation is 0.18 seconds, which is 0.5% based on the minimum of 36.13 seconds. The deviations result from basic processes of Windows XP, the properties of processed data, and a variable network situation, which causes deviations because the browser is used as an object. The performance improvement in terms of CPU is 19.30% through comparison of the values of the two arithmetic means. Accordingly, the improvement ranges from 0 to 19.30% whose lower and upper bounds correspond to simultaneous failures of 100% and 0% of primary and first secondary replicas, respectively. Therefore, the expected improvement is the arithmetic mean of 9.65% assuming the ratio of simultaneous failures of primary and first secondary replicas is 50% over all objects.

The results for maximum usage of memory are shown in Fig. 5.

The peak of memory usage ranges from 47.69 to 51.25 MB for the type of 3 replicas. The arithmetic mean is 48.69 MB and the standard deviation is 1.46 MB, which is 3% based on the minimum of 47.69 MB. On the other hand, the peak of memory

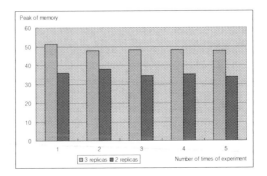

Fig. 5. Maximum usage of memory in MB

usage ranges from 33.84 to 37.88 MB for the type of 2 replicas. The arithmetic mean is 35.47 MB and the standard deviation is 1.57 MB, which is 4.6% based on the minimum of 33.84 MB. The deviations result from the same causes as in case of CPU described earlier. The performance improvement in terms of memory is 27.14% through comparison of the values of the two arithmetic means. Accordingly, the improvement ranges from 0 to 27.14% whose lower and upper bounds correspond to simultaneous failures of 100% and 0% of primary and first secondary replicas respectively. Therefore, the expected improvement is the arithmetic mean of 13.57% assuming the ratio of simultaneous failures of primary and first secondary replicas is 50% over all objects. The simulation has been performed with another URL of www.nia.or.kr inside Korea to investigate how much the properties of processed data and a variable network situation influence the results.

The results for the total CPU time used are shown in Fig. 6.

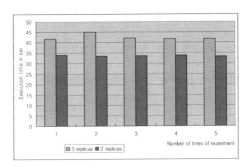

Fig. 6. Total time of CPU use in sec.

The total CPU time used ranges from 41.73 to 45.03 seconds for the type of 3 replicas. The arithmetic mean is 42.48 seconds. On the other hand, the total time of CPU use ranges from 33.38 to 34.06 seconds for the type of 2 replicas. The arithmetic mean is 33.68 seconds. The performance improvement in terms of CPU is 20.71%

through comparison of the values of the two arithmetic means. Accordingly, the improvement ranges from 0 to 20.71%. Therefore, the expected improvement is 10.36% which is 0.71% higher than that with the previous URL.

The results for maximum memory usage are shown in Fig. 7.

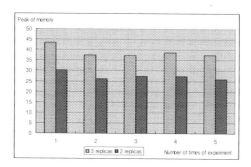

Fig. 7. Maximum usage of memory in MB

The peak of memory usage ranges from 37.24 to 43.41 MB for the type of 3 replicas. The arithmetic mean is 38.86 MB. On the other hand, the peak of memory usage ranges from 25.94 to 30.31 MB for the type of 2 replicas. The arithmetic mean is 27.42 MB. The performance improvement in terms of memory is 29.43% through comparison of the values of the two arithmetic means. Accordingly, the improvement ranges from 0 to 29.43%. Therefore, the expected improvement is 14.72% which is 1.15% higher than that with the previous URL. To sum up, the influence of the properties of processed data and a variable network situation on the ratio of performance improvement in terms of CPU and memory is very low.

5 Conclusion

The ITS can be deployed based on FT-CORBA efficiently considering heterogeneous and real time properties of it. However, improvement is needed to enhance performance of the ITS based on FT-CORBA because it requires additional uses of CPU and memory for object redundancy. This paper has proposed an architecture to adjust the number of object replicas autonomously and dynamically with an autonomous agent of the FTAgent. In the future, additional research is needed to optimize the number of object replicas in real environment of ITS as follows.

Firstly, the FTAgent can improve performance of its own over time by learning from statistical data related to recovery of replicas by objects such as the interval to check failures and their frequency, which means improvement of the line (14) through (19) of the algorithm. The learning/adaptation is one of properties in the context of intelligent agents lacking in the FTAgent. Accordingly, an additional module and its DB are needed to manage statistical latencies between recovered time from previous faults and occurrence time of current ones of replicas by objects for last one hour. They enable the FTAgent to be an intelligent agent which is defined as one that is capable of flexible autonomous action to meet its design objectives. If the latencies are longer than statistical minimum time for real time services of transport information, the FTAgent can reduce the number of related replicas to 2 in case of this paper. The statistical latencies must be updated continuously with data of recent one month. Secondly, the size of the DB maintained by the FTAgent has to be studied experimentally as well which is the record of failures for one month in this paper. It will be decided according to the characteristics of transportation information which generates in real time. Recently, real time information of transportation has been being provided to people through various interactive media, e.g., mobile phone, PDA, internet, and IPTV. The proposed architecture will be useful to meet those increasing demands. It can be applied to other FT-CORBA based systems as well because the ITS is a composite one to have properties of most applications.

References

1. Balasubramanian, J., Gokhale, A., Schmidt, D.C., Wang, N.: Towards Middleware for Fault-tolerance in Distributed Real-time and Embedded Systems. In: Meier, R., Terzis, S. (eds.) DAIS 2008. LNCS, vol. 5053, pp. 72–85. Springer, Heidelberg (2008)
2. County of Los Angeles Department of Public Works, http://www.ladpw.org/TNL/ITS/IENWeb/index.cfm
3. Ezhilchelvan, P.D., Macedo, R.A., Shrivastava, S.K.: Newtop: A Fault-Tolerant Group Communication Protocol. In: 15th International Conference on Distributed Computing Systems, pp. 296–306 (1995)
4. FatihAkay, M., Katsinis, C.: Performance improvement of parallel programs on a broadcast-based distributed shared memory multiprocessor by simulation. Simulation Modelling Practice and Theory 16(3), 347–349 (2008)
5. Felber, P., Guerraoui, R., Schiper, A.: The Implementation of a CORBA Object Group Service. Theory and Practice of Object Systems (TAPOS) 4(2), 93–105 (1998)
6. Felber, P., Narasimhan, P.: Experiences, Approaches and Challenges in building Fault-tolerant CORBA Systems. Transactions of Computers 54(5), 497–511 (2004)
7. Franklin, S., Graesser, A.: Is it an Agent, or just a Program?: A Taxonomy for Autonomous Agents. In: Jennings, N.R., Wooldridge, M.J., Müller, J.P. (eds.) ECAI-WS 1996 and ATAL 1996. LNCS, vol. 1193, p. 25. Springer, Heidelberg (1997)
8. Gokhale, A., Natarajan, B., Schmidt, D.C., Cross, J.: Towards Real-time Fault-Tolerant CORBA Middleware. Cluster Computing: the Journal on Networks, Software, and Applications Special Issue on Dependable Distributed Systems 7(4), 15 (2004)
9. Guan, C.C., Li, S.L.: Architecture of traffic.smart. In: 8th World Congress on ITS, pp. 2–5. ITS America, Washington (2001)
10. IONA Technologies, http://www.iona.com/
11. Lee, J.K.: IICS: Integrated Information & Communication Systems. Journal of Civil Aviation Promotion 23, 71–80 (2000)
12. Moser, L.E., Melliar-Smith, P.M., Agarwal, D.A., Budhia, R.K., Lingley-Papadopoulos, C.A.: Totem: A Fault-Tolerant Multicast Group Communication System. Communications of the ACM 39(4), 54–63 (1996)
13. Nagi, K., Lockemann, P.: Implementation Model for Agents with Layered Architecture in a Transactional Database Environment. In: 1st Int. Bi-Conference Workshop on Agent Oriented Information Systems (AOIS), pp. 2–3 (1999)

14. Narasimhan, P., Dumitras, T.A., Paulos, A.M., Pertet, S.M., Reverte, C.F., Slember, J.G., Srivastava, D.: MEAD: support for Real-Time Fault-Tolerant Corba. Concurrency and Computation: Practice and Experience 17(12), 1533–1544 (2005)
15. Natarajan, B., Gokhale, A., Yajnik, S.: DOORS: Towards High-performance Fault Tolerant CORBA. In: 2nd Distributed Applications and Objects (DOA) conference, pp. 1–2. IEEE, Los Alamitos (2000)
16. Natarajan, B., Gokhale, A., Yajnik, S., Schmidt, D.C.: Applying Patterns to Improve the Performance of Fault Tolerant CORBA. In: 7th International Conference on High Performance Computing, pp. 11–12. ACM/IEEE (2000)
17. Object Management Group: Fault Tolerant CORBA. CORBA Version 3.0.3 (2004)
18. Saha, I., Mukhopadhyay, D., Banerjee, S.: Designing Reliable Architecture For Stateful Fault Tolerance. In: 7th International Conference on Parallel and Distributed Computing, Applications and Technologies (PDCAT 2006), p. 545. IEEE Computer Society, Washington (2006)
19. Vaysburd, A., Birman, K.: The Maestro approach to Building Reliable Interoperable Distributed Applications with Multiple Execution Styles. Theory and Practice of Object Systems (TAPOS) 4(2), 73–80 (1998)
20. Vehicle Information and Communication System Center: VICS units reach 3.2 million. Technical report, 8th World Congress on ITS (2001)

A Platform for LifeEvent Development in a eGovernment Environment: The PLEDGE Project

Luis Álvarez Sabucedo, Luis Anido Rifón, and Ruben Míguez Pérez

Telematics Engineering Department,
Universidade de Vigo
{lsabucedo,lanido,rmiguez}@det.uvigo.es

Abstract. Providing eGovernment solutions is becoming a matter of great importance for governments all over the world. In order to meet the special requirements of this sort of projects, several attempts have been and are currently developed. This papers proposes its own approach that takes advantage of resources derived from the use of Semantics and from an artifact, deeply discussed on the paper, defined as LifeEvent. On the basis of these premises, a entire software platform is described and a prototype developed, as shown in the paper. Also some conclusions and hints for future projects in the scope are provided.

Keywords: eGovernment, semantics, interoperability.

1 Introduction

In these last years, the provision of a holistic software framework for the fulfillment of needs from citizens turned out to be a very demanding area. Governments from all over the world have been devoting a large amount of resources to provide solutions capable of meeting the citizen needs.

Several definitions of eGovernment can be provided, from UN[1][1], from the World Bank[2][2] or from EU[3][3]. Upon its review, it is clear that eGovernment is not just a simple replacement of technology to provide a 24/7 service. Indeed, provision of eGovernment solutions involves a huge effort in re-engineering all processes involved in the public service to place the citizen at the center of the process. As a matter of fact, this technology forces PAs to re-orient and improve services by positioning the citizen at the center of all provided operations. These

---

[1] The use of information and communication technology (ICT) and its application by the government for the provision of information and basic public services to the people.

[2] Refers to the use by government agencies of information technologies that have the ability to transform relations with citizens, businesses, and other arms of government.

[3] Is about using the tools and systems made possible by Information and Communication Technologies (ICTs) to provide better public services to citizens and businesses.

Y.A. Feldman, D. Kraft, and T. Kuflik (Eds.): NGITS 2009, LNCS 5831, pp. 146–157, 2009.
© Springer-Verlag Berlin Heidelberg 2009

services should be, whenever possible, an end-to-end transaction in order to achieve a one-stop digital administration.

Nowadays, as a result of the large amount of resources, a great number of project are implemented or in process to be completed. Upon its review, limitations in deployed solutions can be find out and some concrete problems to overcome (see Section 2) can be pointed out. This papers presents an comprehensive review of them and propose a platform that actually take into account these concerns to provide is own solution. In particular, problems related to the discovery and accessibility (in terms of software architecture) for services in the domain are addressed.

Bearing these concepts in mind, the use of a modeling tool called LifeEvent (LE hereafter) is proposed. This artifact models services from the point of view of the citizen in terms of their need. These LEs implies the use of Administrative Services (AS hereafter) to actually carry out services. The main goal of this proposal is to support in a simple manner from the point of view of the citizen searches and queries about domain services, as shown on Section 3.

In order to increase the possibilities of the system regarding automatization and interoperability issues, semantics is bring into scene. Nowadays, among the scientific community, semantics is usually considered the enabler technology to develop this sort of solutions. This technology offers us a new set of tools and capacities which have not been completely explored yet (see Section 4). Finally, some brief notes on the deployment of the system are provided (see Section 5). Finally, some conclusions are presented to the reader in Section 6.

2   Analysis of the Problem and Motivation

As mentioned above, eGovernment can be considered as the next generation of service paradigm for the civil service. ICTs are surely bound to play a paramount role in this task. Also, we can notice a strong demand of eGovernment solutions. This demand has two components. On the one hand, there is a growing demand of public services from citizens who are willing to take advantage of ICT based solutions. On the other hand, there is a latent demand due to new laws in most countries compelling them to cover a wider range of services with telematic support.

Anyhow, a common feature in all solutions is the use of the Web support to deliver contents. Acting as a front-end solution for citizen, these Web pages offer an access to services. Nevertheless, as shown below there is still a long road ahead in order to achieve the desired quality level. In a not very formal manner, we can classify the available services on web portals according to the following categorization[4], from simpler to more complex:

- Presence. Just information to access on the web such as forms to download, web maps, . . .
- Information about the town. It is possible to access information updated periodically about specific topics related the PA: transport, minutes from meetings in the council hall, map of the city, useful telephone numbers, . . .

- Interaction. Support for assistance from the PA in a digital manner: email support for information and some operations.
- Complex operations. Real operations supported in a holistic manner: support for personal folder, tax payments, security features, ...
- Support for political management. Services related to eDemocracy are available: support for polls, interaction with the agenda of the meeting in the council hall, ...

The vast majority of eGovernment solutions are presented to citizens by means of Web portals. The highest functioning Web portals show a complete system integration across agencies, whereas portals with the lowest level of functionality provide little more than access to forms and static pieces of information. High-functioning portals create a true one-stop service for citizens. In particular, usability, customization, openness and transparency represent key aspects of portal functionalities. Regretfully, upon the review of web portals[5], several drawbacks can be outlined. In our case, we are going to focus on problems related to locating services. This is not a simple task. When looking for a particular service in the web site of a PA, it is not a trivial task to find the proper place where the service is held. This is due to the wide variety of classifications for services, mechanisms for its invocations, visual interfaces and even problems such as finding out beforehand if the administration is responsible for the wished service.

By means of the review of several already deployed eGovernment platforms[5], notable difficulties for citizens have been unveiled. Two main shortcomings are present in current-fashion solutions from the point of view of the citizen:

- It is not a simple task to discover which is the right service for a particular situation in the life of a citizen. This problem may be due to the variety of administrations that can be in charge (e.g., requesting a grant can depend on different administrations), to the difficulties in finding the services (e.g., the use of some searchers can be difficult if the user is not involved in administrative issues) or to the lack of information in the expected format.
- Administrations have not foreseen mechanisms to orchestrate services. In those situations where the citizen needs support from two or more administrations, he/she is on his/her own.

In general, no special treatment for locating the desired services is provided. This way, there is no a common or interoperable service to deal with the addressed goal: searching the most suitable service for a citizen from several administrations.

3   LifeEvents to Model Services

Taking into account the above mentioned considerations, a LE-based approach is proposed to deliver advanced tools in locating services. The very first goal is to model services from the point of view of the citizen and no longer from the perspective of the PA and its internal procedures.

Therefore, LEs can be defined in the scope of this proposal as "any particular situation in the life of a citizen that must be dealt with and requires assistance, support or license from PAs". Those situations are considered from the point of view of the citizen and only include operations meaningful for the citizen. We can consider as LEs, in the frame of this proposal, situations such as paying a fine, getting married, moving, losing one's wallet ...

Taking the analysis of the domain, already existing tools and the first evolution of the prototype as a basis, we are in a position to identify some relevant fields in the definition of LE:

- Task. Title for the considered operation.
- Description. High level description of the desired operation expressed in natural terms from the point of view of the citizen. This field will support human-driven selection procedure to finally choose the right LE.
- Input documents. All operations carried out by the administration require some input document. The citizen must provide, at least, a signed form in order to invoke the operation. This field states required documents to undertake the LE. We do not include simple forms or similar documents in this category since only meaningful documents for the citizen are taken into account at this point.
- Output documents. Of course, as a result of any performed operation, the PA in charge, whatever this may be, must provide an output expressed in terms of the ontology. This information will be put together into one or several documents. The content of the output will vary, ranging from the expected document (e.g., a certification, a license ...) to information about potential failures in getting the expected document. This field will state the legal documents that will be issued to the citizen in response to his/her request. On the basis of these fields searches can be described.
- Scope. We must identify the scope in which we want an operation (local, national, international...) to be recognized.
- Area. The particular area concerning the LE: education, taxes, health care, retirement, ...
- Security Conditions. The conditions for the security mechanism involved in the whole process, such as identification of both parties, cipher methods, etc. Using this field, the citizen can express the minimum conditions required to deal with his/her personal documents.
- Version. LEs can be modified and changed from one version to another. This information, which aims at tracking versions and providing coherent management of LEs, is not to be displayed to the citizen.

Note that no PA is attached to a LE as a LE may involve several PAs and the citizen may no be aware of this situation.

3.1 Administrative Services

If we take the definition of LE as a basis, a single LE can –and normally does– involve different operations in different administrations. For our proposal it is quite important to model these situations also. As already mentioned, LEs just model the need but for an eventual invocation, it is necessary to know which administrative services are required to really have a service. In order to model those interactions, another accessory artifact is introduced in the system: the Administrative Services. PAs can, and normally do, provide support to perform simple operations held directly by them. These operations may involve a more or less complex back-office mechanism but, in any case, they are invoked and fulfilled in a certain PA. Therefore, the fulfillment of a LE could involve several ASs in different PAs. ASs include the following data:

– Title. Brief name for the AS.
– Description. This information will be used in the description of ASs as components of LEs.
– Deadline. The maximum span of time for the response from the PA before the operation is considered approved/dismissed. This information is used to estimate how long an entire LE may last.
– Public Administration. Information about the PA; it is responsible for the execution of the PA and it is used to decide about the scope of the operation
– Law. Information about the supporting laws involved in the current AS. This could allow citizens to gather further information or support from government or lawyers in case of any eventuality.
– Input and Output documents. As in the case of LEs, these documents will provide the information about the required inputs to the system and the expected outputs. This information plays a main role in making decisions about the orchestration of ASs.
– Internal operations. Although an AS is easily executed in a PA, this process may have a number of steps that will be used to inform the citizen about the evolution of his/her request.
– Related ASs. This field informs about ASs that can usually fall together. These ASs will be the most likely to be coordinated.

4   Semantic Support

The semantic web has emerged as a new promising technology aimed at addressing information instead of data, i.e., it enables software agents to treat data in a meaningful manner. Making this possible would allow new mechanisms to operate on a higher level of abstraction. Also, by means of this technology, it is possible to express knowledge in a formal and interoperable way. These features will support the provision of the support claimed in this paper.

4.1   Semantics

The "semantic", as an IT researching field, was born in the early 2000's. In May 2001, Sir Tim Berners-Lee published the foundational article presenting the semantic web to the world[6].

"The Semantic Web will bring structure to the meaningful content of Web pages, creating an environment where software agents roaming from page to page can readily carry out sophisticated tasks for users".

The main goal of this idea is to make machines capable of understanding the information within the web. This feature will allow them to make more complex interactions without the need of human support. To accomplish this ambitious goal, a long evolution on the technological side has taken place during these last years. In this context, ontologies are a key element. In the literature, several definitions or approximations to the concept of ontology are provided. A rather suitable definition for ontology may be[7]:

An ontology is a formal, explicit specification of a shared conceptualization of a domain of interest.

By means of this definition, we are addressing an ontology as a support to present abstract information about a certain domain in a concrete way by means of a machine-understandable data format.

To express an ontology in a formal manner, different languages[8] are at our disposal. OWL (Ontology Web Language)[9] is the W3C Recommendation intended to provide a fully functional way to express ontologies. To make different levels of complexity possible, OWL provides different sublanguages with increasing expressivity: OWL Lite, OWL DL and OWL Full. By using OWL, we are addressing a standard, solid and interoperable platform for the provision of this solution.

Another semantic tool that serves our propose quite nicely is Semantic Web Rule Language (SWRL)[10]. This is a proposal for a rule-language in the Semantic Web broadly accepted. By mean of logic rules, it enables the generation of new knowledge not existing previously in the system.

4.2   Applying Semantics

Up to now, LEs and ASs, as tools to model services in the domain, have been presented with no binding to any technology in particular. For obvious reasons some technical support has to be chosen in order to advance in our proposal. In our case, we decided to make good use of semantics to derive advantages in the system as shown below.

Semantics, in the present case, is presented by means of a modular system of ontologies that describes all entities involved in the process.

Several acceptable options to develop an ontology that may fit our needs are available, such as Cyc[11], Kactus[12], Methontology[13], On-To-Knowledge[14], etc. The chosen option in our case was Methontology[13]. The main reasons for this election were good support with software tools (such as Protégé[15] or OntoEdit[16]), platform-independence; in addition, it is recommended by FIPA[17] and it has been tested in several large scale projects.

The resulting ontology includes the definition of all the meaningful data in the system. Therefore, this ontology must be known and shared by all agents. It is mainly made of some basic classes that characterize the ground.

- citizens. As the system final users, they are expected to be able to invoke all services available and interact with information in the system.
- documents. They provide the legal support for all operations in the system. Security concerns must be deployed in their management, since they compel agents to respect the contents of the documents.
- LifeEvents. These elements model the need of citizens and they are the drivers towards the service completion provided by the Digital Lawyers. Certainly, their definition includes already mentioned fields.
- AdministrativeServices. This class holds the properties to express all the knowledge about ASs as they are defined in previous sections.

In this ontology, we have reused previously proposed metadata. For example, in the task of defining the citizen, one of the main classes in the system, FOAF (Friend of a Friend)[18] has been reused. Also, metadata from European standardization bodies has been reused to mark documents in the system, in particular, CWA 14860[19] from CEN[20]. This is part of a general philosophy leading toward the maximum possible agreement and reusability both for the ontology and the software based on it (see Figure 1).

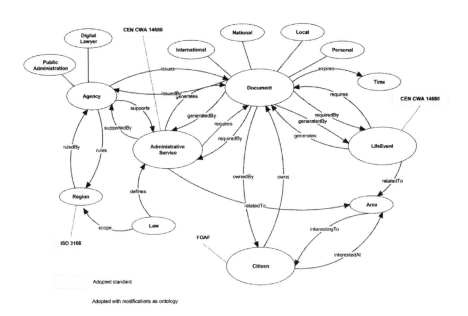

Fig. 1. Ontology to define the knowledge in the system

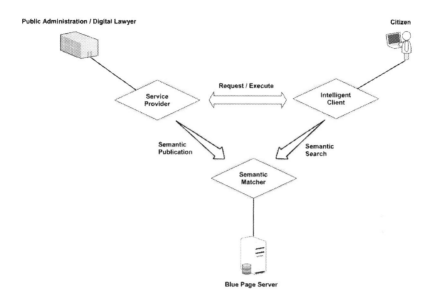

Fig. 2. Agents in the system working cooperatively

5   Implementation

In order to take advantage of the concepts presented in the paper, it was developed a prototype under the support of the PLEDGE project. Therefore, the concepts presented in the paper were used to deploy the actual software support.

The implementation of the entire system involves the inclusion of several agents, namely:

- Citizen. This agent acts on the behalf of citizen supporting task related to the discovery and invocation of services. This agent provides with a front-office application that allow citizens to search/invoke services on any Public Administration engaged in the system.
- PublicAdministration. This agent provides support for the actual execution of services on demand by citizens. Also it supports the integration in the Public Administration itself to keep the track of the execution of the operation and enable the interaction with the citizen once the operation (LE or AS) has been requested.
- Blue Page Server (or BPS). This agent conducts the search for services, LE and AS in our case. This agent with no user interface provides by means of Web Services with facilities for searches from citizens and publication of LE and AS from Public Administrations.

This agents will work cooperatively in order to solve the global goal of the system (see Figure 2).

To carry out the implementation of these agents, authors decided to make use of a Java platform. Main reasons for this are due to its long background in large scale solutions, its capacity to run on nearly any hardware support and the existence of software libraries suitable to our interests. Also, for the implementation of agents, the use of Web Services was a key technology. And under this platform there are available plenty of libraries especially designed for this task. As a result, users of the system, both citizens and Public Administrations, are provided with an user friendly local application (see Figure 3).

In order to implement the semantic features required in this case, the Jena library[21] was selected among other libraries. This library was the chosen option due to the large community of programmers currently working on its development and the support for required semantic operations such as executing queries or merging new instances.

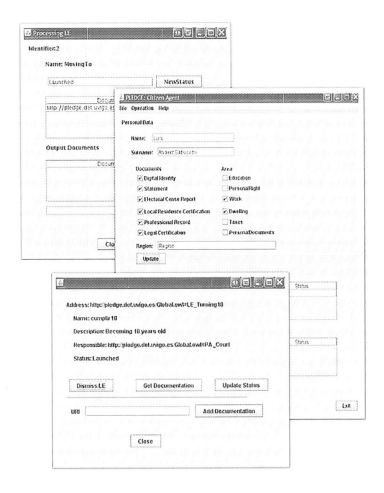

Fig. 3. Screenshots from the prototype

Taking advantage of these artifacts and semantic support available, it is possible to provide support for advanced searching services. Taking into account the profile of the citizen, i.e., areas assigned, documents owned or desired, current location, and so on, it is possible to personalize the possible LE the citizen may be interested at. Also, it is possible to discover which are the best ASs that may be invoked to fulfill a particular LE. Briefly, there are several high-level functions defined to support these functionalities:

- `searchLEbyInputDocument` To locate those LEs that require some document as input.
- `searchLEbyArea` To locate those LEs related to some particular area.
- `searchASbyOutputDocument` To locate those LEs that require some document as input.
- `relatedAS` To put in contact ASs that may be related in terms of area, administration, or expected output.

These operations are in the core of the Blue Page Server, the agent with the higher semantic features in the system. To carry out this functionality, the BPS must gather the information in the pool of LE/AS that are actually store with in its own semantic database. The following snippet shows how all individuals from the ontology with some certain properties can be recalled. In this case, it is illustrated how a LE is found. The mechanism to achieve that consists on executing a SPARQL[22] query with all the information regarding to the desired LE/AS using the following Java code:

```java
// Generating the query in the BPS
String queryString=BPS.generateQuery();

// Generating a ResultSet object
//   with the data
Query query = QueryFactory.
              create(queryString);
QueryExecution qe =
       QueryExecutionFactory.
          create(query, individuals);
ResultSet rs = ResultSetFactory.
       copyResults(qe.execSelect());
qe.close();
```

We would like to outline that even currently the state of the art related to semantic software must be improved quite a lot, the chosen libraries were capable of meeting the requirements of our particular problem.

6   Conclusions

The main goal of this paper is the introduction of an ontology-based support to facilitate tasks related to services locating. By means of LE, several advances have been achieved: services are defined in a more user-friendly manner, a uniform mechanism to characterize them is available, and their localization becomes easier as the semantic annotation provides new tools.

As already mentioned, eGovernment is currently a research field where a lot of effort is being placed. As a result, a large number of efforts and initiatives have arisen. In the literature, we can also find some interesting initiatives that make use, at different levels, of semantics applied to LE-based concepts in some manner: the Finnish Web site Suomi.fi[23], the EIP.AT project[24], the SemanticGov project[25], and the Access-eGov project[26]

Regarding these projects, the presented proposal goes a step further and suggests features not provided in these platforms such as the provision of a LEs pool from different PAs, semantic mechanisms to discover the proper LE, a new and flexible mechanism to compose new services from already existing operations, etc

This proposal takes into consideration limitations and shortcomings from the technological environment, especially from the semantic environment, where a lot of work is yet to be done. Nevertheless, some issues related to the design of an accurate semantic matcher were overcome thanks to the approach selected, where some simplifications are possible.

We would also like to stress the importance of providing with software tools that make easy for both public administrations and citizen to take part in eGovernment solutions. This feature is key in order to get users engaged with this new paradigm of service.

Acknowledgment

This work has been funded by the Ministerio de Educación y Ciencia through the project "Servicios adaptativos para e-learning basados en estndares" (TIN2007-68125-C02-02) and by the Xunta de Galicia, Consellería de Innovación e Industria "SEGREL: Semántica para un eGov Reutilizable en Entornos Locais" (08SIN006322PR).

References

1. United Nations: Benchmarking e-government: A global perspective (2005), http://unpan1.un.org/intradoc/groups/public/documents/un/unpan019207.pdf
2. World Bank: About e-government (2007), http://www.worldbank.org/egov
3. European Commission: Swrl: A semantic web rule language (2009), http://ec.europa.eu/information_society/activities/egovernment/index_en.htm
4. Esteves, J.: Análisis del Desarrollo del Gobierno Electrónico Municipal en España (2008), http://www.sumaq.org/egov/img/publicaciones/3.pdf
5. Papadomichelaki, X., Magoutas, B., Halaris, C., Apostolou, D., Mentzas, G.: A review of quality dimensions in e-government services (2007)

6. Berners-Lee, T., Hendler, J., Lassila, O.: A new form of web content that is meaningful to computers will unleash a revolution of new possibilities. Scientific American (May 2001)
7. Gruber, T.: A translation approach to portable ontology specifications. In: Knowledge Acquisition, pp. 199–220 (1993)
8. Ontological Engineering. Springer, Heidelberg (2003)
9. W3C: Web ontology language (2004), http://www.w3.org/2004/OWL/
10. Horrocks, I., Patel-Schneider, P., et al.: Swrl: A semantic web rule language (2004), http://www.w3.org/Submission/SWRL/
11. Lenat, D.B., Guha, R.V.: Building Large Knowledge-based Systems: Representattion and Inference in the Cyc Project (1990)
12. Schreiber, A., et al.: The KACTUS View on the 'O' World. In: Workshop on Basic Ontological Issues in Knowledge Sharing, IJCAI 1995, pp. 28–37 (1995)
13. Fernández-López, M., Gómez-Pérez, A., Juristo, N.: Methontology: From ontological art towards ontological engineering. In: Symposium on Ontological Art Towards Ontological Engineering of AAAI, pp. 33–40 (1997)
14. Ontoknowledge project: Oil (2005), http://www.ontoknowledge.org/
15. Stanford Medical Informatics: Protege (2007), http://protege.stanford.edu/
16. AIFB, University of Karlsruhe: Ontoedit (2007), http://www.ontoknowledge.org/tools/ontoedit.shtml
17. Fundation for Intelligent Physical Agents: Fipa (2005), http://www.fipa.org/
18. The foaf project (2005), http://www.foaf-project.org/
19. CEN: Dublin Core eGovernment Application Profiles (2004), http://www.cenorm.be/cenorm/businessdomains/businessdomains/isss/cwa/cwa14860.asp
20. CEN: CEN home page (2005), http://www.cenorm.be/cenorm/index.htm
21. Hewlett-Packard: Jena (2005), http://www.hpl.hp.com/semweb/
22. W3C: SPARQL Query Language for RDF (2006), http://www.w3.org/TR/rdf-sparql-query/
23. SW-Suomi.fi - Semanttinen informaatioportaali (2007), http://www.museosuomi.fi/suomifi
24. EIP.AT (2006), http://eip.at
25. The SemanticGov Project (2006), http://www.semantic-gov.org
26. Access-eGov (2006), http://www.accessegov.org/

Online Group Deliberation for the Elicitation of Shared Values to Underpin Decision Making

Faezeh Afshar, Andrew Stranieri, and John Yearwood

Centre for Informatics and Applied Optimization,
University of Ballarat, Victoria, Australia
(f.afshar,a.stranieri,j.yearwood)@ballarat.edu.au

Abstract. Values have been shown to underpin our attitudes, behaviour and motivate our decisions. Values do not exist in isolation but have meaning in relation to other values. However, values are not solely the purview of individuals as communities and organisations have core values implicit in their culture, policies and practices. Values for a group can be determined by a minority in power, derived by algorithmically merging values each group member holds, or set by deliberative consensus. The elicitation of values for the group by deliberation is likely to lead to widespread acceptance of values arrived at, however enticing individuals to engage in face to face discussion about values has been found to be very difficult. We present an online deliberative communication approach for the anonymous deliberation of values and claim that the framework has the elements required for the elicitation of shared values.

Keywords: Shared Core Values, Deliberation, Computer-mediated communication.

1 Introduction

Values an individual holds including honesty, freedom and fairness have been shown to underpin attitudes, behaviour and relationships, and motivate decisions [1]. Rapid global changes in recent decades have made the importance of values more obvious than ever before. Numerous authors [2, 3, 4, 5, 6] link many global issues such as environmental problems with a lack of emphasis on values in many areas of decision making.

Values are not solely the purview of individuals as communities and organisations have core values implicit in their culture, policies and practices [7]. In organisational contexts some authors [7, 8] have shown that strongly values-based organisations are the most stable, productive and successful. Therefore, the elicitation of core values could be critical for a community of participants that are seeking competitive advantage by addressing the value-based decision making needs of our time. In a similar vein, Barrett [9] argues that the alignment between the stated values and beliefs of an individual or group and their actions and behaviours leads to authenticity and integrity which he regards as a solid foundation for trust.

Y.A. Feldman, D. Kraft, and T. Kuflik (Eds.): NGITS 2009, LNCS 5831, pp. 158–168, 2009.
© Springer-Verlag Berlin Heidelberg 2009

According to Dolan and Garcia [2], the setting of organizational values is largely believed to be the role of management. It is quite unrealistic to expect that values for an organisation will be aligned with those held by all individuals, given the complexity and diversity inherent in most communities and organisations. Further, according to Hofstede [10] individual values are typically set at a young age and only slowly change. Values set by management will differ from individual values however a greater degree of acceptance of organisational values can be expected if individuals have an opportunity to participate in the process of setting values, rather than have them imposed by management. The participation by individuals in the process of setting organisational values ensures that misalignment between individual and organisational values can be kept to minimum and that the reasons for the organisational values are well understood.

Typically, values shared by a group are derived by eliciting values ranked by importance that each individual holds using an instrument such as the Schwartz Value Survey (SVQ) [1]. The individual ranked lists are merged using any one of a number of algorithms advanced for this purpose in order to generate a shared ranked list of values. However, this approach to merging individual rankings avoids discussions that could lead to a deeper consensus and greater acceptance of shared values.

This paper contends that another way to reach shared values is to enable proper deliberation, explanation and reasoning to lead to a consensus on core values based on a shared understanding. Open deliberation on the ideal value rankings for the organisation is likely to lead to a greater appreciation of the range of individual values and also a deeper understanding of reasons underpinning each choice.

Open deliberation about values is not straight forward. In face to face settings, individuals have been found to be reluctant to openly divulge their values, particularly if they are dramatically misaligned with those of their corporation [11]. Individual values are deeply personal and open discussion about values can be intimidating and require a very high level of inter-personal trust.

The ConSULT approach [12, 13, 14, 15] provides an online, anonymous framework for many participants to express their ideas, views, explanations and reasons and ultimately reach a decision by consensus. This forum is well suited to discussions about values. Reaching a consensus decision in ConSULT occurs through the articulation of all reasoning for and against all propositions in a deliberative, anonymous argumentation process that allows free participation and contribution in a cooperative online environment.

In the next section an overview of values is presented before describing the ConSULT approach in more detail.

2 An Overview of Values

Schwartz and Bilsky [16] defined values as "concepts or beliefs, about desirable end states or behaviors, that transcend specific situations, guide selection or evaluation of behaviour and events, and are ordered by relative importance" (p. 551). Schwartz [1] considers the concept of values as central to understanding the principles that guide individuals, institutions and organizations. Groups and individuals' values are used to understand important issues within cross-cultural management, politics, psychology, and belief systems [17, 18].

Mitroff and Linstone [19] considered that organizations can not only be viewed from technical perspectives but also from personal, ethical and aesthetic perspectives. Collins and Porras [20] have identified organization values as central to achieve high performance. Hall [21] positions human values at the core of organisations which influence its structure and culture. These core values define the philosophy and reasons underlying the existence, objectives, operations and decision making in organizations.

Buchko [22] presents two main perspectives in organizational values. One views organizational values as the enduring systems of belief in an organization. From this view, organizational values take precedence over individual values. Consequently, organizations often hire individuals who are perceived to fit with the organization's values. The other perspective views organizational leaders as vital elements in the development and implementation of organizational values and values-based behaviours. For example Dolan and Garcia [2] indicate that often, just one or two leaders' values create a culture in an organization.

Nohria and Ghoshal [23] indicate that the presence of a high degree of shared values across individuals in an organization improves performance and the proper definition and the use of such values can lead to many positive outcomes. Buchko [22] goes further in suggesting that the enforcement of formal structures and policies could be replaced by the use of shared values as a means of influencing individual behavior. The importance of shared values was noted by Ouchi [24] as a central characteristic in creating a strong organization culture. Morgan and Hunt [25] indicate that whilst shared values can help towards harmony, understanding and shared vision; the lack of shared values often are the bases of misinterpretation, misunderstanding and conflict. Blanchard [26] identified the effectiveness and the importance of organizational shared values as one of the essential requirements in the practice of management and leadership in organizations.

While there is ample acknowledgement that shared values lead to positive outcomes for an organization, there are also indications that an individual benefits from an alignment of personal values with organizational values. Fischer and Smith [27] noted that individuals interpret how they are perceived, valued and treated by management based on their own value structures. Consequently, their perception of justice is based on the extent to which their treatment by management corresponds with their own individual values. Hall [21] suggests that individuals are happier and more productive when their values are in alignment with the espoused organizational values.

Shared values have also been identified as important for decision making. Stoddard and Fern [28] relate the effectiveness of group decision making to values. Wong [29] explains how personal values of individuals underlie the decision-making process and Fritzsche [30] relate decision making ethics to values. Courtney [31] maintains that effective decision-making and management in organizations are based on the way problems are perceived and suggests that organizations should recognize the humanity aspects of its members. He argues that besides technical aspects in the decision making processes, equal consideration should also be incorporated for ethical perspectives. This consideration is especially important when the problem is of a social nature.

A relatively small set of around 56 values has been found to be universally regarded as core values [1, 32]. All individuals acknowledge the same universal values as important but differ in the way in which the values are preferred, typically represented as a ranked list. One of the pioneers of values research, Rokeach [33] regarded values as enduring beliefs that make a specific mode of conduct or end-state of existence personally or socially preferable to alternative ones. He presented his Rokeach Value Survey (RVS) for individuals to rank values to attain their value profile. While, Rokeach [33] distinguished a way to explain how values are organized as value hierarchies; Schwartz [1] observed that considering values as simple hierarchies can ignore their interrelatedness and further distinguished a way to describe values as structures which include both compatible and conflicting values. Schwartz [1] made strong assumptions about the value types and with his structural relationships extended the theory of Rokeach [33]. He specified 56 values and their descriptions which makes his SVS (Schwartz Value Survey) inventory.

Schwartz [1] derived a typology of the different contents of values from a need for individuals within the reality of their social contexts to cope with three universal requirements of human existence: individual needs as biological organisms, requisites of coordinated social interaction and requirements for the smooth functioning and survival of groups.

These three requirements led to a model of universal structures of values that distinguishes among 10 value types based on their motivational goals. The ten basic values types are universalism, self-direction, stimulation, hedonism, achievement, power, security, conformity, tradition and benevolence. Schwartz [1] and [34] listed the value types described in terms of their central goal, and identified the single values that primarily represent them. Schwartz and Sagiv [35] found that 8 out of 10 value types form distinct regions in a multidimensional space, and 2 types, that are highly related in the theory, are intermixed. This led to the arrangement of values shown in Fig. 1.

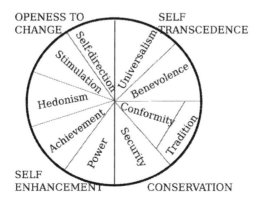

Fig. 1. Theoretical model of structure of relations among 10 value types [34]

As shown in Fig. 1 value types form a range of related motivational continuum that can be presented in a circular structure; except for *Tradition* which shares its continuum section adjacent to *Conformity*. In his structure, the closer the values in either direction around the circle the more similarity of underlying motivations, and their distance indicate the conflict between underlying motivations. Schwartz [1] posits that the behaviour connected to each value type has psychological, practical, and social consequences that may be compatible or in conflict with another type. He proposed that while some value types are compatible and motivated by certain value types, others could generate strong psychological and social conflicts. For example, the practice of *Achievement* values are often in conflict with the practice of *Benevolence* values, because to seek personal success is likely to get in the way of the actions that are required to help the welfare of others. [36]

Schwartz has developed two instruments to study people's value systems: Schwartz Value Survey (SVS) and Portrait Value Questionnaire (PVQ). The SVS was found to be difficult for respondents to use for finding personal values and led to the development of the PVQ. The PVQ consists of 40 short verbal third person portraits of different people in which the values of respondents are inferred from their self-reported similarity to those described people [37]. Schwartz et al. [37] noted that respondents typically describe SVS as an intellectually challenging task that demands deliberation and weighing of the relative importance of values. Respondents to the PVQ treat it as a simple task, quick to respond (7 to 10 minutes complete) and report no difficulty in making judgments and rarely ask any questions.

The SVS and PVQ are instruments for the elicitation of individual values that have been empirically validated across many settings and cultures. Less work has been done toward the elicitation of organizational values. This is discussed next.

3   Deliberation to Elicit Organisational Values

Most of the methods to elicit organizational core values are based on the aggregation of many individual values. For example Schwartz used SVS to find collective values in most of his studies by the aggregation of all the individual ones. Although this may be sufficient for uncovering the aggregation of current values; the results could be too fragmented and not sufficiently focused to give a group a deliberated, clear and satisfactory outcome as a base for values to be adopted by everyone in the consequent decision makings and behaviours.

In his Discourse Ethics, Habermas [38] demands consideration of the viewpoints of all people affected by decisions. He further stresses that the participation in a discourse should be conducted with full awareness of the other people's perspectives and interpretations. Therefore, a consensus process needs to be held in an environment that provides equal opportunities for every individual in the group to share their opinions, suggestions, assertions and supporting or opposing reasons to propositions.

Boje and Murnighan [39] show that independent of any kind of outcome, acceptance and agreement increases group unity and participants' commitment to group decisions. In order that everyone at least accepts the outcome, there should be no one who has any disagreement that has not been valued and considered in the outcome decision. Nemeth and Wachtler [40] indicate that unresolved concerns of minorities often promote resentment, conflict and lack of motivation to act willingly on the implementation of the final decision. Yet, simply allowing a concern to be expressed often helps a resolution of these negative consequences.

Accordingly, we propose the use of deliberation as a means of identifying the shared values in organizations. Deliberation is a dialogue type whereby the contribution of each participant in decision-making is considered valid; deliberation encourages reasons and underlying assumptions and understandings to accompany points of view, and obliges respect and consideration of the reasons of others. The goal of this process is to deepen one's own point of view with that of the others towards enhancing one's own decision.

ConSULT [12] is a computer-mediated, deliberative approach that a community of participants can use to better understand and facilitate the elicitation of multiple perspectives and ultimately reach a consensus decision. In this approach, various techniques from three fields of study; Argumentation, Delphi, and Voting have been integrated to provide a computer mediated deliberation platform for group discussions that culminate in consensus and shared understanding.

The ConSULT framework draws on an argumentation structure presented by Toulmin [41] and two well known approaches that have been used to enhance the quality of decisions that result from group decision making; IBIS (Issue Based Information Systems) [42] and Delphi [43, 44]. Key features of the ConSULT approach include anonymity, absence of a facilitator, asynchrony and the use of the Borda Count voting algorithm [45]. Other algorithms for comparing participant's ratings to determine majority views include pairwise comparisons however the the Borda Count is computationally simple and has been shown to satisfy Arrow's fairness criteria [46] [47].

There are times when individuals themselves keep quiet and do not express their opinions due to some social or psychological factors such as avoiding the stress caused by confrontation, shyness or lack of status or other barriers to freedom of participation in face-to-face group decision-making processes as indicated by [42]. Nanschild [11] discusses how some of her research subjects were not participating in the discussions of values. She found it illuminating that given an expectation for public servants to be able to discuss values in the workplace; this was not the case and in the contrary, her research subjects felt very uncomfortable to engage in a discussion on values. The lack of expression of disagreement in face-to-face meetings could cause a false assumption of unanimity and agreement. This assumption could lead to an illusionary consensus.

Consensus decisions need to be reached by group interaction that promotes participation, gives everyone power to express their opinions, encourages people to listen to each other, considers concerns rather than ignoring them and eliminates the possibility of the choice of only one or a few individuals determining a proposal. This environment should encourage cooperation, trust, respect, unity of purpose, and detachment. Participants in a ConSULT process advance through three phases in determining consensus and a shared understanding. [14]

1. Deliberation and collection of all contributions in the discussion.
2. Anonymous and independent evaluation and voting by participants
3. Re-evaluation through an iterative voting process only this time with the knowledge of collective trends in previous rounds.

Fig. 2 and Fig. 3 shows screen shots from an actual study reported by [12]. Fig. 2 shows a screen shot of participants discussing an issue (Teaching conditions at the School of X are appalling) and Fig. 3 shows a screen shot of a participant's vote for suggestions and reasons for a suggestion.

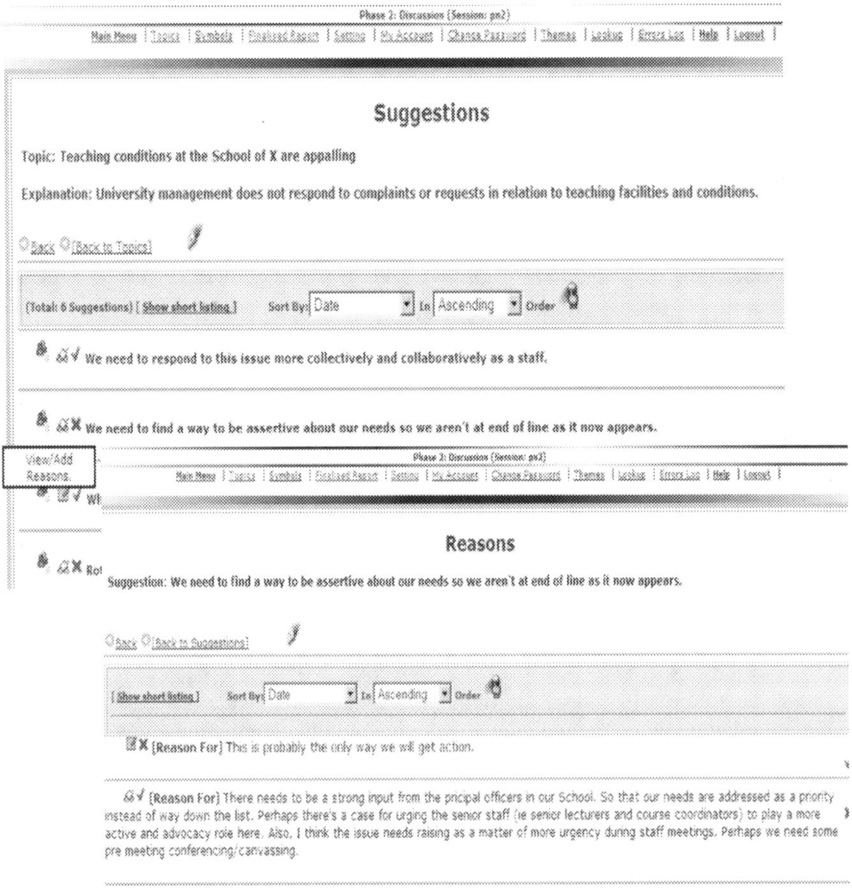

Fig. 2. An issue, its description and deliberation of a suggestion

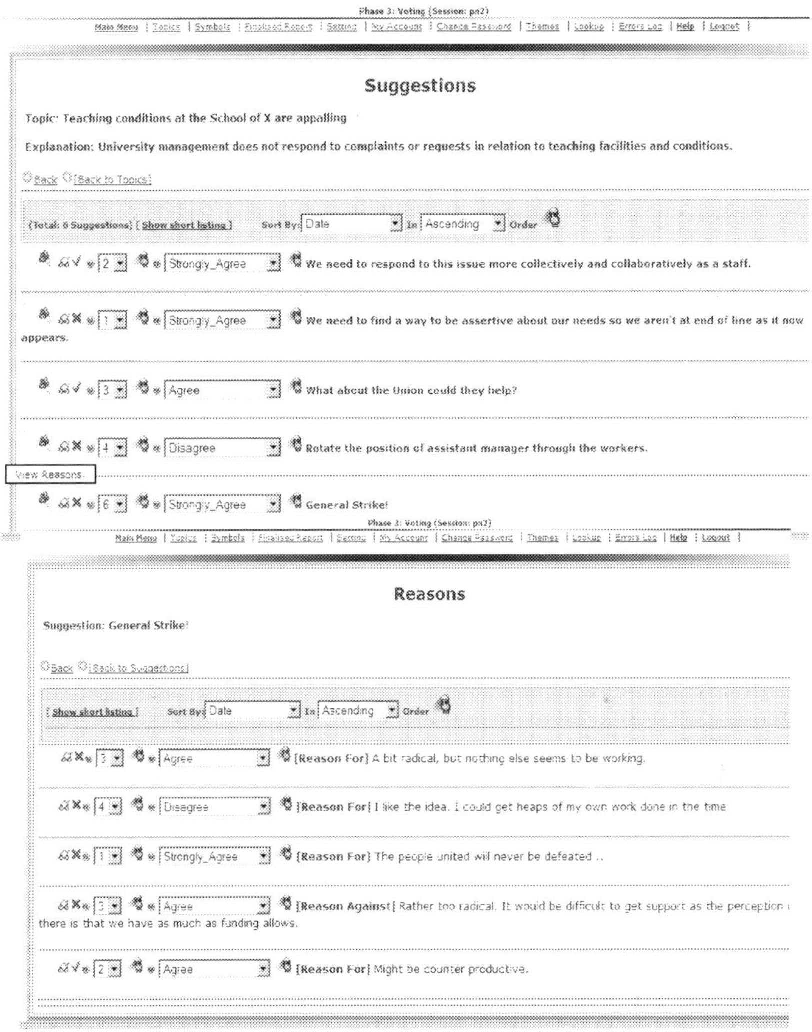

Fig. 3. Voting for Suggestions and Reasons

4   Conclusion

Values are deeply rooted, relatively consistent beliefs which guide and interpret one's opinions, actions and behaviours and impact on all aspects of one's life. They influence how one understands and interact with their reality. The interdependencies and rapid changes in the context of society have made the importance of values more pressing than ever before [48, 49, 50]. With the advancement of technology, scientific discoveries can give us tools to better provide for our physical needs, but only positive values can tell us how to use those tools beneficially. The problems of the environment are example of symptoms of the kind of values that are underpinning organizational economic and social decision makings.

Different methods have been used in order to elicit organizational core values. Most of the methods are based on the aggregation of values by individuals selected from a ranked list of values. We advance deliberative democracy as a means to identify the common or shared values in organizations. This paper proposes an online, anonymous approach to facilitate deliberation to elicit core values may be achieved by considering individual values. It also suggests the use of technology as a basis for multiple perspective decision making support. This paper could contribute towards developing organizational core values with the elicited shared values of employees. The elicited values could be analysed by well validated value instruments to find out their impact in achieving the most acceptable and beneficial outcomes for the organization, their society and environment.

References

1. Schwartz, S.H.: Universals in the content and structure of values: Theoretical advances and empirical tests in 20 countries. In: Zanna, M.P. (ed.) Advances in experimental social psychology, pp. 1–65. Academic Press, San Diego (1992)
2. Dolan, S.L., Garcia, S.: Managing by values: Cultural redesign for strategic organizational change at the dawn of the twenty-first century (2002)
3. Elworthy, S., Holder, J.: Environmental Protection: Texts and Materials. Butterworths (2005)
4. Parson, E.A., Fisher-Vanden, K.: Integrated Assessment Models of Global Climate Change. Annual Review of Energy and the Environment 22, 589–628 (1997)
5. Patz, J.A., Gibbs, H.K., Foley, J.A., Rogers, J.V., Smith, K.R.: Climate Change and Global Health: Quantifying a Growing Ethical Crisis. Ecohealth 4, 397–405 (2007)
6. United Nations: World Commission on Environment and Development. Our Common Future. Oxford University Press, New York (1987)
7. Kaplan, R.S., Norton, D.P.: The Tangible Power of Intangible Assets: Measuring the Strategic Readiness of Intangible Assets. In: Harvard Business Review On Point Collection, pp. 14–29. Harvard Business School Publishing Corporation, Boston (2004)
8. Bragdon, J.H.: Book Excerpt: Profit for life: how capitalism excels. Reflections 7, 55–62 (2006)
9. Barrett, R.: Building a Values-Driven Organization: A Whole System Approach to Cultural Transformation. Butterworth-Heinemann, London (2006)
10. Hofstede, G.: Attitudes, values and organizational culture: Disentangling the concepts. Organization Studies 19, 477–492 (1998)
11. Nanschild, D.: Explicating Values as the Epicentre of Public Sector Leadership for New Times: Putting the 'V' Factor back into the Australian Public Service Values Framework. The International Journal of Knowledge, Culture and Change Management 8, 131–140 (2007)
12. Afshar, F.: A Computer-Mediated Framework to Facilitate Group Consensus Based on a Shared Understanding - ConSULT. University of Ballarat, Ballarat (2004)
13. Afshar, F., Yearwood, J., Stranieri, A.: Capturing Consensus Knowledge from Multiple Experts. In: Bramer, M.A., Preece, A., Coenen, F. (eds.) The Twenty-Second SGAI International Conference on Knowledge Based Systems & Applied Artificial Intelligence, pp. 253–265. Springer, London (2002)

14. Afshar, F., Yearwood, J., Stranieri, A.: A Tool for Assisting Group Decision Making for Consensus Outcomes in Organizations. In: Voges, K.E., Pope, N.K.L. (eds.) Computational Intelligence for Organization Managers, Business Applications and Computational Intelligence, ch. 16. Idea Group Publishing, Hershey (2006)
15. Afshar, F., Straneri, A., Yearwood, J.: Toward Computer Mediated Elicitation of a Community's Core Values for Sustainable Decision Making. In: ACKMIDS 2008 (2008)
16. Schwartz, S.H., Bilsky, W.: Toward a universal psychological structure of human values. Journal of Personality and Social Psychology 53, 550–562 (1987)
17. Miller, J., Bersoff, D.M., Harwood, R.L.: Perceptions of Social Responsibilities in India and the United States: Moral Imperatives or Personal Decisions? Journal of Personality and Social Psychology 58, 33–47 (1990)
18. Hofstede, G.: Cultural Constraints in Management Theories. The Academy of Management Executive 7, 81–94 (1993)
19. Mitroff, I.I., Linstone, H.A.: The Unbounded Mind: Breaking the Chains of Traditional Business Thinking. Oxford University Press, New York (1993)
20. Collins, J.C., Porras, J.I.: Built to Last: Successful Habits of Visionary Companies. Harper Business, New York (1994)
21. Hall, B.: Values Shift: A Guide to Personal & Organizational Transformation. Twin Lights Publishing, Rockport (1995)
22. Buchko, A.A.: The effect of leadership on values-based management. Leadership & Organization Development Journal 28, 36–50 (2007)
23. Nohria, N., Ghoshal, S.: Differentiated fit and shared values: Alternatives for managing headquarters-subsidiary relations. Strategic Management Journal 15, 491–502 (1994)
24. Ouchi, W.G.: Markets, bureaucracies, and clans. Administrative Science Quarterly 25, 129–141 (1980)
25. Morgan, R.M., Hunt, S.D.: The Commitment- trust Theory of Relationship Marketing. Journal of Marketing 58, 20–38 (1994)
26. Blanchard, K.: Managing by values. Executive Excellence 14, 20 (1997)
27. Fischer, R., Smith, P.B.: Values and organizational justice: Performance and seniority-based allocation criteria in UK and Germany. Journal of Cross-Cultural Psychology 35, 669–688 (2004)
28. Stoddard, J.E., Fern, E.F.: Buying Group Choice: The Effect of Individual Group Member's Prior Decision Frame. Psychology & Marketing 19, 59–90 (2002)
29. Wong, B.: Understanding Stakeholder Values as a Means of Dealing with Stakeholder Conflicts. Software Quality Journal 13, 429–445 (2005)
30. Fritzsche, D.J.: Personal Values: Potential Keys to Ethical Decision Making. Journal of Business Ethics 14, 909–922 (1995)
31. Courtney, J.F.: Decision Making and Knowledge Management in Inquiring Organizations: A New Decision-Making Paradigm for DSS. Decision Support Systems Special Issue on Knowledge Management 31, 17–38 (2001)
32. Rokeach, M.: From individual to institutional values: With special reference to the values of science. In: Rokeach, M. (ed.) Understanding human values, pp. 47–70. Free Press, New York (1979)
33. Rokeach, M.: The Nature of Human Values. Free Press, New York (1973)
34. Basic Human Values: Theory, Methods, and Applications -An Overview, http://www.yourmorals.org/ schwartz.2006.basic%20human%20values.pdf

35. Schwartz, S.: Value priorities and behavior: Applying a theory of integrated value systems. In: Seligman, C., Olson, J.M., Zanna, M.P. (eds.) The Ontario Symposium on Personality and Social Psychology: Values, pp. 1–24. Erlbaum, Hillsdale (1996)
36. Ros, M., Schwartz, S.H., Surkiss, S.: Basic Individual Values, Work Values, and the Meaning of Work. Blackwell Pub., Oxford (1999)
37. Schwartz, S.H., Melech, G., Lehmann, A., Burgess, S., Harris, M., Owens, V.E.: Extending the cross-cultural validity of the theory of basic human values with a different method of measurement. Journal of Cross-Cultural Psychology 32, 519–542 (2001)
38. Habermas, J.: Discourse Ethics. In: Moral Consciousness and Communicative Action. MIT Press, Cambridge (1993)
39. Boje, D.M., Murnighan, J.K.: Group confidence pressures in iterative decisions. Management Science 28, 1187–1197 (1982)
40. Nemeth, C.J., Wachtler, J.: Creative problem solving as a result of majority vs. minority influence. European Journal of Social Psychology 13, 45–55 (1983)
41. Toulmin, S.E.: The uses of argument. Cambridge University Press, Cambridge (1958)
42. Computer Based Delphi Processes,
    http://eies.nJit.edu/~turoff/Papers/delphi3.html
43. Mitroff, I., Turoff, M.: Philosophical and Methodological Foundations of Delphi. In: Linstone, H.A., Turoff, M. (eds.) The Delphi Method, Techniques and Applications, pp. 17–37. Addison-Wesley Publication Company, London (1975)
44. Linstone, H.A.: The Delphi Technique. In: Fowles, J. (ed.) Handbook of Future Research, pp. 273–300. Greenwood Press, Westport (1978)
45. Hwang, C.L., Lin, M.J.: Group decision making under multiple criteria: methods and applications. Springer, Berlin (1987)
46. Arrow, K.J.: Social choice and individual values. Wiley, New York (1963)
47. Saari, D.G.: Basic geometry of voting. Springer, Berlin (1995)
48. Buchanan, B.: Assessing human values (1997)
49. Harmes, R.: The Management Myth: Exploring the Essence of Future Organisations. Business and Professional Press, Sydney (1994)
50. Cameron, K.S.: Ethics, virtuousness, and constant change: How to lead with unyielding integrity. In: Tichy, N.M., McGill, A.R. (eds.) The Ethical Challenge, pp. 185–194. Jossey-Bass, San Francisco (2003)

Author Index